The Incredible Shrinking Bee

The Incredible Shrinking Bee

Insects as Models for Microelectromechanical Devices

James V. Lawry
California Academy of Sciences, USA

ICP
Imperial College Press

Published by

Imperial College Press
57 Shelton Street
Covent Garden
London WC2H 9HE

Distributed by

World Scientific Publishing Co. Pte. Ltd.
5 Toh Tuck Link, Singapore 596224
USA office: 27 Warren Street, Suite 401-402, Hackensack, NJ 07601
UK office: 57 Shelton Street, Covent Garden, London WC2H 9HE

Library of Congress Cataloging-in-Publication Data
Lawry, James V.
The incredible shrinking bee : a biominiaturization farrago / by James V. Lawry.
p. cm.
Includes bibliographical references.
ISBN 1-86094-585-6 (alk. paper)
1. Bees--Physiology. 2. Size judgment. I. Title.

QL563.L39 2006
573.1'15799--dc22

2005052142

British Library Cataloguing-in-Publication Data
A catalogue record for this book is available from the British Library.

Typeset by Stallion Press
Email: enquiries@stallionpress.com

Printed in Singapore by World Scientific Printers (S) Pte Ltd

To Muddy who believes
she was put on Earth
to be my mother.
I came with bees.

"I have endeavoured in this Ghostly little book to raise the Ghost of an Idea, which shall not put my readers out of humour with themselves, with each other, with the season or with me."

Charles Dickens

Encouraged through the edges and vertices of a highly complex network over time and space, with this book I have endeavored to keep Dickens's end in view. I acknowledge the inspiration and help received of mentors, colleagues and friends who have given me the courage to follow my heart and intuition and whose ideas and nurturing contributed to this book in so many ways: John P. Coghlan, Ed Dike, Ivan Barker, Francis Dealtry, Louis Epstein, Barbara Mensing, Robert Miller, Robert Orr, Ed Ross, Earl Harold, George Mardikian, Rolf Bolin, Don Abbott, Donald Kennedy, James Adams, Welton Lee, Vicki and Dan Pearce, Adrian Horridge, Mick Callan, Herbert Macgregor, Mike Lavarack, Ian Monie, Jack DeGroot, Fran Ganong, Alan Mines, Rita Giacaman, Howard Fields, Julius Krevans, Roy Steinberg, Antje Dettmer Lawry, Alan Mines, Bob Drewes, Conrad Vial, Celia Lovell and especially Carol Sweig, my wife, who still puts up with me.

And lastly, I thank Joy Quek and Aileen Tan of World Scientific (Singapore) and Lenore Betts at ICP (London). After all, these dear folks published this book.

Jim Lawry

CONTENTS

GLOSSARY

ADSORPTION: Attachment and bulk transfer of a substance from a fluid onto a surface.

ATOMIC FORCE MICROSCOPY (AFM): A scanning-probe microscopy that maps the topography of an interface by scanning a force sensor over the interface.

ATP: Adenosine triphosphate is a high-energy molecule containing phosphate used as a cellular energy currency to store transport and release energy within the cell.

BLOOD PRESSURE: The hydrostatic pressure of blood within the arteries. In vertebrates, the top number or the systolic pressure, reflects pressure in the arteries when the heart contracts; the bottom number or the diastolic pressure, is the pressure in the arteries while the heart fills between beats.

BOTTOM-UP: Building up a complex object from its smaller components.

CAD: Computer aided design employs hardware and software for design and technical drawing permitting simulations of designs without constructing physical prototypes. Using CAD, components of designs may be reused and standard components and assemblies generated automatically.

CARDIAC OUTPUT: The volume of blood pumped in an interval by a heart.

CAVITY TRANSPORT: Transport of substances within a fluid-filled or fluid-lined cavity as compared to transport through a closed circulation.

CMOS: Pronounced sea moss, stands for complementary metal oxide semiconductor.

CHIP: An electronic component or a system of semiconductor components containing one or a group of electronic functions in an integrated circuit on a substrate of silicon.

CHITIN: The tough, protective, semi-transparent substance, mostly of a nitrogen-containing polysaccharide, that is the principle component of arthropod exoskeletons.

CLOSED CIRCULATION: A heart pump and its blood vessels: arteries, arterioles, capillaries, and veins.

COMPLEXITY or COMPLEX SYSTEM: Features include parts couple with each other in a network of small-world or scale free topology; relationships between parts are nonlinear; systems are open and exist in a thermodynamic gradient, dissipate energy, and contain feedback loops. Initial conditions and small perturbations can determine the evolution and history of complex systems, and these systems may be nested and evolve over multiple potential paths. Because boundaries of complex systems may be indeterminate, observers must decide where these lie.

CUTICLE: An insect's hard outer body wall.

DIASTOLE: The resting or filling phase of the cardiac cycle.

EFFICIENCY: The ratio of the work a system performs to the energy the system expends in performing the work.

ELECTRON TRANSPORT CHAIN: Biomolecular machinery within cell membranes and mitochondria that couples flows of electrons to proton pumps converting energy from sugars into ATP.

EMBOLUS: A clot traveling through a blood vessel that lodges and obstructs blood flow downstream from the clot.

EMERGENCE: A large-scale unprecedented and unanticipated group behavior we cannot explain in terms of a system's parts, that appears at many different levels of organization. The number of interactions between a system's components increases with the number of components. Some interactions, however, may be negligible or may cancel each other out or create noise and thereby work against the emergence of system behavior. Emergence may not signify complexity, but emergence may help distinguish which laws and concepts apply to macroscopic scales and which to microscopic ones.

FEEDBACK: Moment-to-moment interaction between a controller and what it controls. In NEGATIVE FEEDBACK a disturbance causes the system to return towards a set point; in POSITIVE FEEDBACK a disturbance compels the system to progress rapidly towards a maximal or minimal state and then remain there.

FRACTAL: Short for fractal dimension. A broken geometric pattern seen in coastlines, terrain, clouds and other patterns exhibiting a simplifying invariance under scale in that geometries repeat themselves at different scales of magnification or resolution and so cannot be represented by a single classical geometric figure; an object whose Hausdorff dimension is not an integer.

GRAPH: A network using the notation V, E meaning a graph having a vertex or node set V and an edge or link set E. Any edge is incident with the two nodes that define it, and the endpoints are said to be adjacent nodes. See Chapter 3: "Beauty Before the Beast," for specific graphs and concepts.

HEMOCOEL: The internal body cavity of an insect containing fluid hemolymph and the internal organs of the body.

HEMOLYMPH: The watery blood of an insect. Does not contain hemoglobin or transport oxygen.

HORMONE: A chemical messenger produced at one location that travels to other parts of the system and attaches to a receptor where it has an effect thereby eliciting responses at a distance from its source. Depending on the receptors and machinery activated, one hormone may perform numerous functions.

LAB-ON-A-CHIP: A microfluidic chip for clinical diagnosis and screening minute quantities of dissolved compounds in which liquids traverse micro-channels.

MALPHIGIAN TUBULES: The excretory organs of an insect.

MARKOV CHAIN: The sequential evolution of a system where, if we are in state i we move to state j with probability Pij after one unit of time.

MASKING: A mask or "photomask" is a square glass plate having a patterned emulsion of metal film on one side. The mask is aligned with a silicon wafer so that the pattern can transfer onto the wafer. Each succeeding mask after the first one must be aligned precisely over the previous pattern. Once a mask has been aligned, the photoresist is exposed through the pattern on the mask using high intensity ultraviolet light.

MEMS: Microscale mechanical devices that combine electronics to make integrated electromechanical actuators for ink jet printer heads, chemical sensors and scanning probe microscopy. Devices range in size from a few micrometers to a few centimeters.

MESOPHYSICS: Quantum behavior of samples or devices containing a very large number of atoms. A continuous but nebulous area bridging between mechanics, biophysics, optics and magnetism where mesoscopic phenomena occur such as in the workings of a macromolecular motor but where mechanical behavior may be understood at the quantum level.

METASYSTEM TRANSITION (MST): A process creating a higher level of organization, the metalevel, often controlling a web of smaller, more subservient sub-systems.

MICROFLUIDICS: Refers to the science behind research and development of microscale devices that contain chambers and tunnels managing nano and picoliter volumes of fluids. Such devices operate in ink jet printer heads, nucleic acid analyzers, labs on chips, and micro-environmental and micro-hematology analyzers.

MICROFLUIDIC CHIP: A plate of silica having narrow channels for moving fluids. Measures chemistries and extracts molecules from solution.

MINIMAL SURFACE: A surface of minimal area having minimal surface tension. The least area spanning a given contour.

MODEL: A simplified hypothetical description of a complex entity or process.

MOORE'S LAW: An empirical trend for the number of circuits per chip to double every eighteen months.

NAD Nicotinamide adenine dinucleotide is a coenzyme undergoing cyclical reduction to NADH+ and oxidation to NAD. Acts as a diffusible substrate for cellular dehydrogenase enymes to provide reducing equivalents for the electron transport chain.

NANOMACHINES: Mechanical devices so small that their parts are single molecules.

NANOMETER: One billionth of a meter. (Units: nm.)

NANOTECHNOLOGY: The technology of building electronic circuits and devices from single atoms and molecules. Devices are less than 100 nanometers in size. Biomolecular nanotechnology utilizes biomolecules as components for machines.

NEMS: Nanoscale mechanical devices.

OPEN CIRCULATION: The invertebrate circulation in which movements of the body assist open pumps to propel blood or hemolymph over and around organs in an open cavity that is usually called a hemocoel but in some groups a coelom.

PARALLEL PROCESSING: An architecture that performs more than one operation at the same time.

PEDICLE: The wasp's waist or connector between the thorax and abdomen of an insect.

PERCOLATION THEORY: Involves a percolation threshold, p, when p defines an average degree of connectivity between arbitrary subunits such as coffee grounds. When p equals zero, all subunits are totally isolated from each other as when the grounds are dry. Wetting the grounds randomly creates connections so p increases. For p less than the percolation threshold only isolated non-spanning clusters exist, and the moist patches remain localized. The percolation threshold is the point at which a spanning cluster first appears. As a spanning cluster forms, coffee starts to drip. When p equals one, all subunits connect to some maximum number of neighboring subunits, and the system percolates in that connected paths traverse the entire system linking one subunit with the next across the spanning cluster. The hemocoel in which fluid accumulates and resorbs may resemble a percolation model.

PERFUSION PRESSURE: The difference between arterial and venous pressures across an organ or capillary bed.

PIPELINING (PIPELINE PROCESSING): Permits simultaneous or parallel processing and refers to overlapping operations by envisioning the sequence of activities within a conceptual pipe wherein all stages of the pipe process simultaneously. As one instruction executes, the next instruction is decoded.

PHOTOLITHOGRAPHY: The process of transferring geometric shapes on a mask to the surface of a silicon wafer to form a chip. Process involves first cleaning the wafer, forming a barrier layer, applying resist, soft baking, aligning the mask, exposure and development followed by hard baking.

POWER DENSITY: A measure for scaling mechanical power. Power is proportional to force times speed that is proportional to area, so power density is proportional to power divided by volume. Because power is the amount of work available this idea expresses work per unit volume of mechanical or electrical energy available.

POWER LAW: A power law relationship exists between two scalars, x and y, when y equals ax^k where a is the constant of proportionality and k, the exponent, are constants. Power laws form straight lines on a log-log graph because taking logs of both sides shows log(y) equals k log(x) + log(a) forming the equation for a line: y equals mx + k. Power laws describe the scaling invariance in many natural systems.

QUANTUM COMPUTER: A computer exploiting quantum mechanical phenomena such as superposition and entanglement.

QUANTUM DOT: An object so small, between two and ten nanometers, that adding or removing an electron results in a single observable change.

QUBIT: The quantum-computing analog to a bit.

RESIST: A material or coating that protects a surface from chemically reacting.

RESOLUTION: The minimum distance between two objects that can be distinguished in microscopy or the minimum spacing between two features that can be fabricated using lithography.

REYNOLDS NUMBER: A dimensionless combination of variables important in analyzing liquid flows when there is a substantial velocity gradient or shear. The number indicates the relative significance of the viscous effect compared with the inertial effect. Compare with Marangoni number. (Note: Balasubramaniam, R. and R. S. Subramanian (2004). Thermocapillary migration of a drop: an exact solution with Newtonian interfacial rheology and stretching/shrinkage of interfacial area elements for small Marangoni numbers. *Ann. N. Y. Acad. Sci.* 1027: 1–8.) During the thermocapillary migration of a drop, the drop's surface undergoes stretching and shrinking as the drop moves. As surfactant molecules adsorb onto the interface of the drop, the surface tension in the interface changes, and this changing area of interface contacting the liquid in turn changes how energy moves in the fluids adjoining this interface. As the drop's interface stretches, the drop's internal energy increases as its area increases. Adjoining fluids must supply this energy and are consequently cooled. Conversely, a shrinking surface element loses energy to the adjacent fluid and consequently warms it. In a moving drop, interfacial areas stretch in the forward parts of the drop and shrink in the rear half causing the surface temperatures of the drop to vary. This interesting model demonstrates a changing convective transport of momentum as values of the Reynolds number vary and reveals one of the numerous intimate relationships between surfaces and the fluids passing over them.

SAFETY FACTOR: Load tolerance.

SCALE FREE NETWORK: A complex network where some nodes are highly connected hubs while most nodes have fewer connections.

SCALING: As the size of a system changes the relationships among its components must adjust so the system continues functioning. Scaling maintains these relationships over a wide range of orders of magnitude. Such self-similarity is called fractal, and the relationship among the variables is described by a fractal dimension or power function.

SCANNING TUNNELING MICROSCOPE: A device in which a sharp conductive tip moves over a conductive surface, typically at a nanometer or less, creating a tunneling current. Commonly one keeps the voltage constant and monitors this current. Raising or lowering the tip above the surface draws an atomic map of the surface revealing its combined topography and electronic properties. The STM can be used to manipulate atoms and molecules on surfaces.

SELF-ASSEMBLY: An integrative mechanism in which components spontaneously assemble while bouncing in a liquid or gas phase permitting stable structures of minimum energy to form. An assembler's process would theoretically not require added input of external information or energy.

SELF-ORGANIZED CRITICALITY (SOC): Embodies the idea that complex behaviors develop spontaneously in certain many-body systems whose dynamics change rapidly. Example: avalanches in a sand pile.

SELF-SIMILAR: Objects or systems where magnified pieces resemble the whole. An example is a cauliflower.

SEMICONDUCTOR: A substance or object having conductive properties between those of a conductor and an insulator.

SETPOINT: The goal for a feedback control system.

STROKE VOLUME: The volume of fluid or blood ejected per stroke from a pump or heart.

SURFACTANT: Short for surface-active agent; a molecule that lowers surface tension.

SYSTOLE: The contraction or pumping phase of a heart's cycle.

TOP-DOWN: Molding, carving and fabricating small materials by using larger objects as tools.

TRACHEAE: The larger tubes of the tracheal system of insects consisting of internal conduits that convey oxygen in air from openings in the cuticle to all parts of the body.

TRANSMURAL PRESSURE: The force across a wall tending to distend or collapse it. Transmural pressure equals pressure inside the wall minus the pressure outside the wall.

VLSI: Very large scale integration permitting more than one hundred thousand transistors on a chip.

PREFACE

> 'No honeycomb is built without a bee
> adding circle to circle, cell to cell,
> the wax and honey of a mausoleum,
> this round dome proves its maker is alive.'
>
> Robert Lowell

At times our restless imaginations seek insights unencumbered by what we definitely know. We observe nature then read and wander in imaginary spaces. We then return to nature to see if some of the things we imagined are there and whether there or not, we pose questions. We realize everything we see connects with so many things we do not.

This book is like this.

We know so little about the inner workings of the smallest insects. But insect ideas may serve us best when we design our smallest tools and machines. This book is not a complete study. So please do not approach it as such. It is a farrago, a potpourri of related ideas and some of their ramifications. To experts in the disciplines touched upon, coverage may appear elementary, but because this book tries to pose new questions to entomologists, systems biologists, physiologists, mathematicians, engineers and computer scientists and anybody else having interest, its scope must be general. I hope everyone will read for a big overview of how bees move things around in their bodies and afterwards are sufficiently motivated to have their own thoughts. How much of the imaginary does our real world exclude anyway?

Origins

This book arose from a paper, 'Insects Separate Diffusing Particles in Parallel,' presented at the *2001 Fourth International Conference on Modeling and Simulation of Microsystems*. Discussions of how the circulations of insects and other invertebrates transfer heat, mass and momentum within microfluids through phase interfaces of complicated geometries at sizes important for developing our own compact energy and chemical systems suggested to me that many groups searching for similar grails did not talk to each other. Systems modelers, computer scientists, mathematicians and engineers were largely unaware of how insects did the things smaller and better that they were trying to do. On the other hand, entomologists and biological "types" were conspicuous by their absence from this meeting. Hence this book.

James Lawry
California Academy of Sciences, San Francisco

Chapter one

WHAT'S IN THIS BOOK

Introductory Note

I hope many people from numerous disciplines will leaf through and perhaps even read this book. Because mathematics and especially equations discourage so many readers, I present quantitative ideas verbally, but for more formal coverage, I annotate several seminal papers and general references in the references section. I also omit tables, graphs and many figures, because more up-to-date and deeper examples are online. I hope *The Incredible Shrinking Bee* stimulates diverse reading, heated discussions, and many new ideas.

Why Study Bees?

This book shows how we may use bees and other insects as models for our smallest machines. Insects and spiders are all smaller than our smallest 'stand alone' devices. Some can pass through a needle's eye. Because insects circulate their blood differently than we do ours, insects can be small. Bees are master miniaturists. Might we ever create machines as small and as competent as bees?

Vector Competency

Apart from suggesting ideas for micro-machines, understanding transport of fluids within insects may have enormous health implications. ARBO or arthropod-borne diseases kill people and animals. All major groups of pathogens have evolved into arthropod vectors, and at least six groups of arthropods evolved blood feeding. We have today arthropod-borne viruses, tick-borne rickettsial diseases, and mosquito-borne malaria and yellow fever. How well an insect transmits disease is its vector competency. Vector competency depends upon biochemical, physical, genetic and environmental factors. Interfering with a vector's competence is one way to control the vector as well as spread of the disease.

Role of Circulation in Disease

Insects and ticks counter a host's blood clotting mechanism, and to be successfully transmitted, parasites must overcome biochemical and physical barriers. Usually arthropods ingest pathogens from a vertebrate host during a blood meal. Pathogens then emerge from the blood meal in the gut of the arthropod and then pass through the wall of the gut into the cavity of the circulatory system: the hemocoel. Within the cavity of the hemocoel, hemolymph or insect blood circulates the pathogens throughout the body and to the salivary glands. The pathogens invade the salivary glands, and the arthropod's next blood meal transmits the virus into its victim.

For example, for viral infections such as dengue, RNA viruses persist in nature because blood-eating arthropods keep passing viruses to new hosts. The viruses multiply in the hosts' blood to very high numbers, so when the next arthropod vector ingests them with its meal, the viruses then multiply in the tissues of the arthropod before passing on to a new vertebrate host.

A better-known example is malaria. Soon after a mosquito ingests a blood meal, male and female malarial gametes emerge and join to form zygotes within the blood in the gut of the mosquito. After about two days, the zygotes penetrate the wall of the midgut of the mosquito where in about a week they become oocysts. Inside the oocysts, the parasites multiply into thousands of sporozoites. Then on about day twelve, the sporozoites enter the mosquito's hemocoel where the parasites flow with the hemolymph to invade the salivary glands, so that during the next blood meal, the mosquito injects malarial sporozoites into its new host.

Were we to understand hemolymph's circulation more deeply, we might physically block parasites in vectors and reduce vector competency. Malaria kills more than one million people a year. Most are in Africa: pregnant women and children under five. Despite new drugs and better mosquito nets, deaths may be increasing because of breakdowns in public health systems. Moreover, mosquitoes eventually develop resistance to DDT and all other chemical pesticides used on them so far (Ref: Arthropod Vectors).

Insect Circulation Differs From Ours

Insect blood does not carry oxygen. Instead, insects use separate conduits called tracheae to transport oxygen. These little tubes convey atmospheric oxygen directly to cells and muscles through tiny portholes in their skeletons. Keeping oxygen transport separate from insect blood sets a maximum upper limit to how large insects can be; the largest were about thirteen inches. Compartmentalized circulatory and respiratory systems permit insects to be very small.

Millions of Years of Research and Development

Insects arose at least 350 million years ago, and over deep geological time the trial and error processes of evolution created the diverse bodies of present day insects. Evolution adapted insects to their changing world. Bodies of today's insects compared with those of fossils are miniaturized and more efficient. We know little about the lives of the earliest insects, as we know so little about life in the Devonian Period of the Paleozoic Era. The earliest insects were large. They crawled, could not fly, and were adapted to cold. Even though the world warmed, some ancient traits survived. For example, adult midges of one species can still walk on ice with their bodies at minus eighteen degrees centigrade. Evolutionary R and D provide today's scientists and engineers with a plethora of highly varied, self-contained, fuzzy black boxes that are cheap to produce, versatile and robust. If we learn what's inside these boxes and how they work, we may plagiarize insect ideas, modify and adapt the insect plan and build smaller machines.

Prototypes

To build small, we must first see the world and learn about it anew 'through insect eyes.' As mammals, we often think 'mammalianly' and imagine that other animals do things the way we do them, so when we design our devices, we may first try to make them too big. We then may try to shrink our larger machines. But in their small world, insects face different 'issues.' For example gravity poses little problem; insects fall without injury and can land upside down on ceilings, but water is a menace. Surface tension traps unwary drinkers requiring insects to have long legs and special mouthparts to avoid the dangers of drink. Insects have mastered small.

Masters of Small

Each bee is a compact packaged web of connected and closely interwoven subsystems. Each subsystem fits perfectly together with all her others, as insects do not tolerate extraneous redundancy. Extra weight only increases need for energy, and fuel is expensive. As knowledgeable wilderness survivalists, bees reduce weight wherever possible. Unlike many of our human miniature systems, bees remain unattached by wires or tubes to batteries or fuel reservoirs. Instead, bees carry their fuel close to their motors. Without being tethered, bees fly, crawl and behave socially. Having their skeletons on the outside leaves large unobstructed spaces inside their bodies.

Inside, a bee has a built-in pilot, motors driving powerful wings and legs, and a circulating supply of fuel. Her body specializes in circulating fluids and gases through pumps and tubes. As her respiratory tubes deliver oxygen directly to where it is needed, she may slowly circulate her blood that not only distributes fuel but also dissipates wastes, heat and carbon dioxide.

A bee coordinates her inside activities with her outside world. Eyes, chemical receptors and hairs sensitive to contact and pressure direct her nervous system to create patterns of electrical messages. These messages pass to muscles and glands, signaling them to contract or squirt. Although a few integration centers such as her brain and other neural centers coordinate some of her activities, she mostly employs many local control centers that are not hard-wired to a central processor to tell her body parts what to do when her life changes.

Bee Fluid Dynamics

If we learn how blood moves inside bees, we should be able to produce similar flows within our own fluid-filled systems built around other ideas stolen from insect circulations. Our model bees could then show us how best to create and exploit control and delivery systems in our smallest devices.

How Can Bees Be So Small?

This book introduces the bee's circulation that can be more useful than the vertebrate system as a model for our small mechanical systems. The insect body is versatile. It shrinks easily, but ours is built too big to shrink very far. Insects have evolved a way to circulate blood in their bodies using a system that also functions in larger arthropods, like crabs and lobsters, but also, not only does a bee's circulation work when shrunk, but it works best in the smallest forms. In fact, the smaller an insect is, the better its circulation appears to work. Before we show why bees can be small, let's see why we cannot shrink a woman down as small as a bee.

The Non-incredible Non-shrinking Woman

Like bees we also are pump-tube systems. Unlike bees, however, our pump-tube vertebrate circulations include our lungs. Compared to insects, our vertebrate circulation is a 'closed' system while the bee's circulation is 'open.' A closed circulation has arteries, capillaries and veins. In animals with backbones, a heart-pump squeezes blood through tubes that go first to gills or lungs, and then this blood, now filled with oxygen and still staying within tubes, goes out all over the body before coming back to the heart. A closed tubular system like ours made small enough to supply an insect would not work.

Printer Analogy

Why not? Imagine a printer having plastic tubes and a head of driving pressure to transport ink from a reservoir to where ink is needed. Were this printer shrunk down to the size of a bee, the pump-tube system would stop distributing ink. The bores of the tubes would now be too small, and the resistance the tubes would place on the pump would easily overcome any pressure the pump could generate to force fluid through the narrow bores. Also a shrunken pump would not work. The smaller a pump is, the less force the pump can produce, and the smaller the volume the pump can eject each time it squeezes. So how do insects move fluids around their bodies and still manage to be small?

Organs Float in a Barrel of Blood

Insects and spiders are hollow. Their skeletons are on the outside, and the big space inside them contains their blood. This cavernous space also houses their organs. In some large insects and caterpillars it's like apples in a barrel of blood. The body space is the 'space of blood' or the hemocoel, and the organs, mostly tethered, float in the blood. The hemocoel contains a variable amount of watery fluid together with blood cells and dissolved nutrients. This blood or hemolymph washes over and bathes the insect's organs. How is this possible? Don't insect organs need to breathe? Yes, they do but not through their blood.

Open Pumps Slosh Insect Blood

One or more open pumps circulate blood in the hemocoel. An open circulation lacks capillaries and veins. It's more like a swimming pool hooked to a filter pump that circulates the water. Imagine yourself standing in a pool of water pumping a foot pump one uses to fill air mattresses and plastic balls. You supply the energy for the pump. As your foot rises, the pumping chamber expands drawing water in through the inlet valve. When your foot descends and squeezes the water trapped in the pump, the increasing water pressure closes the inlet valve that prevents back flow from the pump. As the water pressure in the chamber continues to increase, the outlet valve opens. Now water ejects into the pool. If the pool is initially still, but you keep pumping, water begins to circulate slowly. This open pump system lacks tubes for distributing and collecting water, but it circulates the pool.

The Insect Pump: The Dorsal Vessel

The largest pump in the insect hemocoel is the dorsal vessel (Figure 5.2 in Chapter 5). The dorsal vessel is a small-bore tubular pump running along the top of the abdomen, and like the foot pump, the dorsal vessel is open. This heart-pump circulates the watery hemolymph of the hemocoel and operates in concert with one or more, smaller secondary pumps at the bases of wings or legs where these join the body. The accessory pumps direct blood from the hemocoel into the legs and antennae supplying their muscles before it drains back into the hemocoel.

Unidirectional Flow in the Dorsal Vessel

As in the foot-pump, valves in the dorsal vessel encourage flow in one direction, and in bees, this direction is most often from abdomen to head. Blood from the hemocoel enters the dorsal vessel through little holes all along its length. Then moving rings of muscle, that resemble the waves of peristalsis that push food through an intestine, push the blood forward or backward in the dorsal vessel to a new location where blood leaks out again into the hemocoel. Contracting muscles inside the hemocoel that move wings and legs, together with the jiggling of walking and the shakings of pitching and yawing as she walks or flies, enhance our bee's circulation. In this way she circulates her blood but keeps it at a low pressure.

Where Does Her Blood Go?

Blood from the dorsal vessel enters her head and dribbles out over her brain. Blood then flows backwards through the hemocoel of her head and body contacting her muscles, digestive tract and glands. Blood entering the appendages supplies muscles, sense organs and glands of the legs, antennae and wings. Afterwards, blood returns to the dorsal vessel for the next squeeze. Now what does her blood do?

A Marxian Distribution

The blood of insects is not red because it has no hemoglobin and does not carry oxygen. Hemolymph is mostly water and dissolved salts. Cells, digested food products, hormones, wastes, antibodies and even parasites and viruses travel with the hemolymph. As the hemolymph washes over and contacts each organ, each organ of the body takes the things it needs from the blood and also contributes waste substances back into this same blood in a truly Marxian manner: to each organ according to its need; from each organ according to its ability.

Blood Paths Stay Short and Mostly Outside Tubes

Remember there is a huge advantage in having a hemocoel. In the cavity of the hemocoel there is no need for blood going between two places to return to the heart each time in order to be able to go somewhere else. Short distances in the hemocoel and the ability of blood to move to any point in the hemocoel from any other point without going back to the heart or passing through a tube make it possible for substances to distribute and follow shorter paths than the blood in a vertebrate could do.

Low Blood Pressure Promotes Longevity

Most of the time blood in the hemocoel remains outside tubes and at low pressure. Having a low blood pressure when blood volume is so tiny means that small holes in the skeleton do not always cause exsanguinations. In this way a bee can lose a leg, but its blood can still distribute and collect things directly from the organs while she now on five legs limps away.

Hemocoels Adjust to Changes in Volume

During hot periods when water is scarce, insects retain water within their tissues, so there may be less blood inside the hemocoel. Shrinking the volume of circulating blood in a vertebrate might reduce blood pressure so much that the heart would fail.

Low-volume heart failure does not plague insects. In fact, the opposite happens, and exchange improves. With a reduced volume of blood, the hemocoel becomes more efficient. In a smaller fluid volume, dissolved materials must now confine themselves to move within thin, moist films that line the external surfaces of organs and the walls of the cavity. Here receptors that are on or in the surfaces still can respond to changes in concentrations telling the system which materials and metabolites are to be removed or added to these flows. Because there is less fluid now, distances for travel have shortened, so fewer molecules get lost.

Zoom In

At microscopic dimensions, surface contours and asperities project into the hemocoel. What appears to be just a simple wet flat interior surface, at high magnification includes geometries that change as the volume of fluid in the bee falls during desiccation and rises when she feeds. These microscopic interruptions within an organ's contour can determine how blood flows over organs and surfaces. It is what happens when a swiftly flowing creek dwindles in the summer to a trickle; now a greater portion of what flow remains contacts the stones. What at full hydration is a three-dimensional volume in drier times becomes smaller but also thinner and thinner, behaving more and more as would a moist almost two-dimensional film.

Control Points

To move dissolved substances from the hemocoel into and out of cells, micro-quantities of liquid and solid materials, often together, must traverse complicated phase interfaces having complex geometries of their own. Control points for choices occur at these boundary points. Individual local controllers acting simultaneously over all interfaces together throughout the bee determine what the entire system does as a whole.

A Mobile Service Economy

Because some organs are free to move within the hemocoel, they can drift with the flow to places where they are needed. For example, the bee's kidney or the Malpighian Tubules are long flexible tubes. Resembling a hose attached to a pipe at one end, each tubule floats in and is free to move with the hemolymph. The tethered ends attach to and feed into the alimentary canal. Wastes enter the free ends of the tubules and move along each tubule to exit into the alimentary canal. During drier periods, the free ends of these Malpighian Tubules slide around eventually coming to lie in the fluid filled gutters between organs. Services have moved to where they are needed.

Hemocoels Shrink But Still Coordinate

The properties of the changing thickness of the film of hemolymph help us understand how hemocoels can shrink without destroying the bee's coordination of services. Analogy: imagine a series of machines connected by wires in a room. As the room shrinks the machines squeeze closer together, and the wires begin to take up more and more of the decreasing volume, until at very small volumes when the machines are closest together, the spaces between the machines now hold mostly wires.

However, in the hemocoel the organ 'parts' of the system select what they need independently. Organs 'decide' on their own what each needs and what wastes to eliminate without the need of a large heavy brain and a system of nerve-cables to coordinate their behaviors. Because blood moving in the hemocoel transfers heat, mass and momentum to all parts of the bee without need of a central controller, an engineered model of a hemocoel might excel at sorting and distributing different cells and molecules, let's call them scalars, over space and time. A hemocoel model would be incredibly space and time efficient. Why? Because as our modeled hemocoel shrinks, and its circulating volume of hemolymph gets smaller, the distances blood must travel between pick up and delivery points shorten, so that the distribution-collection system becomes more efficient the more it is shrunk.

Why Model Hemocoels?

We might model hemocoels because hemocoels sort and synthesize, and because their control is diffuse, hemocoels avoid point defects. Now let's see what this means.

Sorting

The micro-mechanical subsystems of a hemocoel when modeled or copied into future devices, may eventually be able to sort individual molecules from moving mixtures of different molecules, possibly to provide flows of input materials to arrays of systems oriented in space. Arrays may be a single surface or any number of layers. Our arrays then in turn might process their molecules linearly or in parallel in a deterministic manner. Bees linearly as well as parallel process recycled subunits of chitin during molting to make new cuticle. In our own machines, newly synthesized molecules might be collected together to form or accrete into complex floating or stationary patterns of components that then might unite forming new structures.

Diffuse Control

The local active surfaces of the hemocoel prevent obstructions. Theoretically at least, control of the hemocoel may occur along any boundaries where organs contact the hemolymph. 'Choices' are at any points on these boundaries. Taken together, places where absorption and elimination from organs and cells occur become not only the control points for what goes in and comes out of each organ, but together these choice points form the controller of the entire system as well. By controlling individually their receptors and surfaces, each cell and organ can regulate what it takes in and puts out according to immediate localized need without relying on outside information coming from a distant central processor. Because control of the hemocoel is largely a diffused function spread over so much area, the hemocoel remains controlled and robust even if point blockages develop.

Hemocoel as Microprocessor

It is tempting to think of the hemocoel as a kind of microprocessor. After all, because the hemocoel parallel processes, it is great at what computer scientists call pipelining where all the steps of a sequence operate concurrently. If each stage is time limited, the time saved by the hemocoel in pipelining is proportional to the number of stages. Low level 'instructions' given to the 'hardware' would include the dynamics of the hemolymph. The number of stages completed each second as with a processor is the 'clock rate,' so that a personal computer with a 200-megahertz clock then executes 200 million stages each second. Many computers have more stages and higher clock rates. The hemocoel is perfectly situated to incorporate 'suprascalar' tasks in which it performs more than one set of instructions at each stage. Because the controlled entities of biological systems are most often cells, we can imagine a cache or a small amount of memory kept right at the site of the processor itself. The cache retains the parts of a program that the system most frequently uses, thereby avoiding calling on more distant memory repositories.

Safety Factors

Hemocoels have huge loading tolerances or safety factors that are much larger than those for pump-tube systems. Hemocoels work when very full or almost empty. Were we to incorporate such loading tolerances into our models, we could learn what redundancies and fail-safe mechanisms we might need to prevent failure of our smallest devices. For example, small machines are believed to have many locations within them where a point defect can cause the entire machine to fail. As in the human circulation, an embolus in a coronary or cerebral artery can spell disaster. This default assumption, however, does not usually apply to machines built on the macro-scale, as tolerances are larger, and many macro-machines can continue to function despite numerous point defects. Because blockage of the entire flow through a hemocoel does not occur even with a large number of point blockages, we might use macro-machine assumptions in modeling this small system, so that we might tolerate even a high density of point defects in our designs. What should become important for our hemocoel models are the shapes or configurations of the surfaces inside the hemocoel and their fluid interfaces.

New Models and New Control Systems

Understanding insect fluid dynamics so as to be able to model a hemocoel might lend novel insight into creating potentially useful control systems for our smallest devices by reducing the number of centralized controllers and their connecting 'wires.' Remember: too much fluid in a pump-tube system, like our heart and blood vessels, can cause pump failure (congestive heart failure) and lead to overall system failure (death). One human remedy may be to take a diuretic to get rid of the excess fluid or to increase heart or pump function with digitalis, but all the organs of the insect hemocoel during fluid overload continue to function without tampering from without. Many regulators of insect physiology are close to the functions they control because distances inside insects are by definition short. Short distances mean fewer 'wires,' shorter wires, and less weight.

Shrinking Increases Efficiency

So let's imagine that we somehow construct a model hemocoel. Then let's imagine shrinking our model all the way down to the size of a bee, and that we watch what happens. Unlike our own pump-tube system, the smallest hemocoel is most efficient. The hemocoel mechanism eliminates any need for the pumps to maintain a high head of sustained pressure. The system conserves energy as distances are short, and because it requires minimal energy to wash the blood over the organs in a large open system at low pressure. Hemocoels are less prone to interruption by clots. A clot at one location does not stop blood flowing around it to other places. In a pump-tube system a clot in a coronary or cerebral artery is so devastating because once a critical artery blocks, there are almost no alternate routes for blood to follow.

Our Bottom Line: Hemocoels Adapt to Changes That Would Block Closed Pump Tube Systems

Remember, hemocoels can adapt easily to changes of volume that occur when the cavity is too full or almost empty, but unlike pump-tube systems, hemocoels are most efficient when fluid volumes are smallest.

Three D Becomes Two D

Think of it this way. Imagine a large hemocoel shrinking smaller and smaller. As the fluid volume grows less and less, at one point this volume eventually becomes just a layer of moisture lining the inside walls of the cavity and the surfaces of the organs. We can see an example with the naked eye if we open the hemocoel of a cicada or locust in the summer. The body cavity is moist inside, but there's not much free fluid. Diffusion of substances within an almost two-dimensional plane film permits maximal control of diffusing substances, as these now can never get lost in the volume of the film. Given infinite time, a randomly moving particle on a plane contacts every point of the plane. In an almost two-dimensional layer of fluid the probability of a molecule moving randomly by Brownian motion from one point to any other in the surface approaches one. However, in a three-dimensional volume of fluid things are very different. There are so many ways for particles to get lost. Not only may a particle diffuse along all four compass directions, but it can go up and down as well. Now the probability is closer to a third for a particle leaving a specific place and arriving at another by diffusion alone. Two-thirds of the particles never arrive. In three dimensions unless a particle leaves a place and arrives at its destination in just a few steps, it never arrives. However, pumping of the dorsal vessel and jiggling from walking and flying may add convection to diffusion, so having a heart makes all our chances better.

Now You Have It

So now you know what *The Incredible Shrinking Bee* is about. Chapter 1 is the shortest, simplest statement of it. In Chapter 2, we compare bees with our micro-mechanical devices such as they are.

Chapter two

BEES AND DEVICES

Overview

Bees and devices are machines. Both couple domains of energy. Both extract energy from the environment, convert it into other forms of energy and liberate heat. Both bees and devices must obey the laws of physics. But bees differ from devices in that bees can reinvent themselves and adjust to environmental changes and can reproduce. Before we can understand enough to copy nature's ways, we must rethink what we mean by emergence, complexity, scale and energy reserves. Chapter 2 reviews what we can do now: chips, NEMS and MEMS, and we compare our efforts with bees. How close to nature are we? How far must we go?

Why Bees?

Insects are successful and adaptable, incredibly so. There are five to at least ten million species of insects but only forty-two thousand species of vertebrates. Because insects are so small we know little about them. Unknown insect lives can be incredibly rich sources for our devices. Insects are robust. They function superbly in all climates at atmospheric pressures. One indicator of insect success: global insect biomass is 10^{12}th kg worldwide, and insect numbers may be 10^{18}th assuming each insect has a mean body mass of 1 mg. In contrast, the biomass of all the world's people is only 2×10^{11}th kg assuming $\sim 5 \times 10^{9}$th individuals, each with a mean body mass of 40 kg averaged over a lifespan (Ref: Dudley, 2000). So bees shower us with good ideas. Bee designs are lean and optimal as evolution acting over geological time guarantees economy of motion in return for expending minimum energy. But how accessible to us are nature's smaller solutions? How well can our devices match what insects do?

Dissect to Learn?

Can we take bees apart to see how their pieces work together? Yes. But dissection creates difficulties. We cannot isolate a wing and neglect the system of which the wing is just a part. An entire bee not only manufactured the wing but maintains, directs, and even twists and adapts it while flying. We'd like to copy nature, but learning how to transliterate nature's words is not easy. For example, we know how silkworms make silk, but can we make silk?

Silk Dreams

Silk drives engineers to tears. Even though we can sequence silk genes and can splice these into the DNA of goats and bacteria, synthetic silk is still not mass-produced nor is it a high-strength material. Silk's composition is not our problem; we know the chemistry, but how does the silkworm caterpillar weave the strands of silk together to give silk its strength as the fibers pass through the caterpillar's spinneret? Let's face it. We cannot build a spinneret: too many small sticky parts; so it may not be until we can build many rows of artificial spinnerets that coordinate their work together without clogging, that silk will become an industrial material. But all's not lost. We have spin-off from our labors, no pun intended. By mimicking the amide linkages in the chains of polymers as we did with nylon, we made Kevlar. The aramid fiber of Kevlar even contains hydrogen bonds in liquid crystal form similar to a concentrated silk solution that improves Kevlar's strength. But real silk? Not yet. We can even shear fibers as they extrude from our machines, and we can crudely mimic a spider's spinneret to produce a tough fiber called rocket wire, so we are getting there. So we think we comprehend silk-making, its details, and we can even duplicate parts of it, but to perform as a bee would require the smallest machines of the greatest delicacy.

What We Can Do

We perform electronic miracles with silicon. But unlike building microchips and electronics, making bees does not require exotic materials, extreme energies, high temperatures or pressures. Nature rarely uses metals or silicon, so producing bee-like devices from metals and silicon would be astronomically expensive. But we are saved. Bees self-replicate. Can we get a device to do that? For a device to self-replicate, it must contain a blueprint for making copies of itself as well as the synthetic machinery to assemble the requisite materials using energy from the environment. So far no device can. But a DNA computer is on its way. Nonetheless, our devices, crude as they are, have things in common with bees.

DNA Computer

Theory has it a trillion DNA computers housed in a single test tube might perform a billion mathematical operations per second with ninety-nine point eight percent accuracy. In fact, this computer has been theoretically possible for some time. After all, cells manipulate DNA and RNA and encode information much as a computer calculates. Now we have a prototype: a programmable computer in which software, hardware, input and output are DNA. DNA not only would mean a smaller computer; it could compute some operations faster than silicon. A regular computer uses bits that are ones or zeros. Quantum computing utilizes qubits that can be every possible number in between. Knowing certain probabilities suggests how qubits will behave. Like DNA, qubits can link together, and something affecting one can influence the rest, so probabilities can interact. This is a start. We have now shown we can work small. We can also work "biological" without destroying biodegradable DNA, and nucleic acid computers would consume less energy and liberate less heat. Might we find naturally existing machines and then combine their parts? But entire bees? Bees are harder to copy. Bees compute, walk, feed and fly all at the same time. Bees are much more than computers.

Devices, Bees and Philosophy

Devices and bees are emergent systems. And how we define emergence either helps or hinders our thinking. Many difficulties hide within the fuzzy, deep, philosophical concepts of emergence, scale, complexity, and energy storage. We must rethink what we think we know and have taken for granted. Devices and bees are also complex systems obeying obscure laws of a "mesophysics" that we don't yet grasp. Mesophysics describes behaviors we cannot predict or control. Mesophysics reigns over a scale lying in between scales we customarily use. In this range, our usual ideas of size and scale are imprecise and largely unknown. Mesophysics is where many relationships overlap. What goes on in the new Lilliput where nanometers measure the 'small' things and microns and micrograms even millimeters and milligrams describe what's 'bigger'? We don't really know. Emergence, scale, complexity, mesophysics and energy are just some of our difficulties when we try to plagiarize from bees.

Emergent Systems

Bees and devices are more than just the sums of their parts and are, therefore, emergent systems. 'Bee' properties are not just a jumbled mix of wings, legs, fuzz and stinger. A television set is not just a case, wires, a picture tube and transistors. Both are entirely new entities. What do we mean by an emergent system? In short, emergence is a large-scale group behavior of a system we cannot explain in terms of the system's parts.

A Monkey Watches The Red Sox

A monkey watches TV. He is intelligent, so we ask him to find out all he can about the phenomenon of TV. So during a commercial, he pulls the plug, and being intelligent, he reduces the TV to a wiring diagram. Now he thinks he 'understands' TV. He knows where currents, resistances and capacitances are. But if we ask him what happened after he pulled the plug, he does not know that Johnney Damon hit a grand slam and that The Red Sox won the World Series after years of abuse by the New York Yankees. No. Our monkey is in the dark about what happened on TV. He lost important information when he pulled the plug. His choice to intervene and dissect the set prevented his continued observation of TV. His interfering defined what he could learn as well as what he couldn't.

The Monkey's Question?

While 'deconstructing' TV the monkey had to ask himself: At what stage of connection do the parts stop being just a group of parts and 'become' TV? When does emergence arise in a system, or how complicated must a system be to be emergent? What is emergence?

Emergence

A major 'rule' for emergent systems is that the parts of a system alone do not make the whole system. We can learn about the parts and how they interconnect, but we also know that something somewhere along the line happens to convert the behavior of individual parts into systemic behavior. When we dissect we destroy what we study and change it to something different, something new. When we intervene we do not know what our effect will be, when it occurs, or the systemic outcome. Had the monkey chosen to disassemble the TV and then count the parts, his understanding would not have advanced, but he would still have lost information.

It is like a doctor asking a patient's sexual history. What the patient thinks of the doctor and his or her 'reasons' for asking the questions determines what the answers are. We may or may not find out what we changed by our intervention. Often what we change is crucial. Is there a better way to get information about emergent systems?

Yes. We can redefine emergence. We can say emergent systems, even with perfect knowledge, will never let us predict all their behaviors. This said; we now can approach our problem differently. If we want to understand a system we must be able to model it. If we can make a working system of our own that behaves as much like the studied system as we would like, then we might say we understand the original system. But as we construct our model, we define what to incorporate in its building so in building we also define the degree of our potential understanding.

In short, to understand a bee, we must make a bee ourselves, a simulated bee. Our simulation *behaves* by showing us 'bee properties.' For each relevant 'bee property' we want our simulated system to have, we must answer questions like: 'Must my simulated wings rotate in flight, or would simpler rigid wings be adequate for my purpose?' "How perfectly must my wing resemble nature's?" In some way, crude or refined, our simulation must do what we think a real bee does. But now we've made progress. A true emergent system is one for which we can optimally predict how it will behave by making a simulation.

What Do We Mean by Understanding?

As our models differ, so will our "understanding." Our models specify the level of our knowledge. Sometimes we employ models without defining them. Here, Richard Feynman tells us of a discussion he had with Fermi:

> 'The calculations were so elaborate it was very difficult. Now, usually I was the expert at this; I could always tell you what the answer was going to look like, or when I got it I could explain why. But this thing was so complicated I couldn't explain why it was like that.
>
> So I told Fermi I was doing this problem, and I started to describe the results. He said, 'Wait, before you tell me the result, let me think. It's going to come out like this (he was right), and it's going to come out like this because of so and so. And there's a perfectly obvious explanation for this.'
>
> Richard Feynman (1945)

Understanding is Relative

Simulating a complex system lets us be as general or specific as our problem requires. Here are three possible responses. After studying a natural bee we might say: 'I understand the rules governing every part of my bee, and I can get my model bee to explain why this and this happens.' Or when we say we understand: 'I understand the rules of bees so well that by watching a bee I can say what it will do without my having to calculate everything.' Or now, our most 'precise' and 'mathematical understanding' of all: 'Analysis allows me to state clearly and precisely what happens through a sequence of states from start to finish to achieve an outcome I define as a "bee outcome." I have a mapping from the space of initial states to the space of outcomes.

If we understand all a bee does, we might design and create a system that does what bees do. To make our model we would have to understand the circumstances necessary and sufficient for each phenomenon to arise in our simulation. So our understanding of our bee system is 'grainy,' but we get to control the graininess. We must always perform some minimum work to predict a phenomenon. We must find the phenomenon, isolate it, study it and then base our model on a set of rules we have copied from nature.

Or John von Neumann: 'The sciences do not try to explain, they hardly even try to interpret, they mainly make models. By a model is meant a mathematical construct which, with the addition of certain verbal interpretations, describes observed phenomena. The justification of such a mathematical construct is solely and precisely that it is expected to work. What does "work" — mean? By trial and error we have determined that those models that work are based on sound physical and mathematics principles gradually wrested from centuries of research. Most of them are characterized by complex partial differential and integral equations that until the advent of computers stood well outside the reach of the engineering and scientific communities.' (Ref: von Neumann, 1962).

> Toto: 'Where are our Sharp Boundaries?'
> Dorothy: 'We're not in Kansas any more.'

Where mesophysics reigns, no sharp line separates emergent and non-emergent parts. We cannot predict exactly where the properties of a group of components become emergent properties. Our inability to predict stems not from our inability to understand but from some inherent property within systems. Might accumulating interactions create "emergence"?

Enter Chaos

Even today's micro-devices are crude when compared with bees. To build insect-sized mechanical systems that also contain sources of energy as to be free of wires and tubes tying them to batteries and reservoirs so they can freely move in micro-scale domains, will require entirely new thinking, new designs and new ways of manufacturing. We are not concerned here with the technical details of how to build bees, but instead, we shall think abstractly about what devices do and how perhaps to avoid silicon and complementary metal oxide semiconductor theory.

Chaos is everywhere. To see chaos, just shrink something. Shrink a device. Will it be more or less efficient in its shrunken state? Because electronics rely on the diffusive motions of electrons, as a device shrinks towards the coherence length for electrons, quantum interference between waves of electrons can dramatically alter what a device does. Capitalizing on quantum effects suggests that nanometer-sized electronic components might find roles within molecular devices. We already know how to modify the motions of electrons inside molecules. But can we build a device as small as a bee?

Shape

Shape determines interactions in time and space, and how systems are connected defines what systems can do. Both bees and many devices lack the uniformity and topological simplicity of electronic circuits, making them extremely difficult to shrink towards chip sizes.

Reproduction

Consider the emergent property of reproduction. To reproduce requires that local and global activities be coordinated. In bees, reproduction depends on complicated and coordinated temporal and anatomical relationships that are orchestrated simultaneously at a genetic, cellular, organ, and organism level of organization. We know from mammals that cells of the pituitary secrete hormones. Sex hormones trigger behavior of sex organs throughout the body, and sexual behavior once started, feeds back to control the pituitary. Insects have analogous hormonal feedback systems. We know little about the rules for interconnecting and timing all the functional components, especially between levels of organization such as cellular and tissue levels. We know more about how cells communicate with cells, but we know so much less about how higher "organismic" patterns become superimposed upon and direct cellular machines all over the body. Learning how a bee controls her own complexity is just one key to increasing our understanding of how to control a new device.

Bees and Devices Integrated Within a Common Manifold

The bee system is exquisitely engineered. Seen more mathematically, a bee as a device is an integrated web of parts together within a common manifold. All parts of a bee share common interwoven properties. These communicate with and between all subsystems of the bee while she performs a division of labor for many other emergent functions, such as mechanical support, muscular work, circulating fluids, reacting to sensory information with coordinated motor responses, and defending her entire systemic self against predation and disease as well as responding to any injury and her needs for growth.

Microfluidic Chip

A device less integrated than a bee by far but suggestive of what bees do is a microfluidic chip. *Labs on chips* isolate diverse particles (proteins, toxins, microorganisms) from dilute solutions of mixed samples. A microfluidic chip is a plate of silica, glass or plastic trenched with narrow channels through which samples flow in tiny streams. Channels may be open or closed. Their walls are coated with substances, monolithic porous polymers, that extract the desired species from solution. Because "receptors" for the substances are only on the walls, only species moving along the walls adhere and are extracted, while most fluid passes unimpeded along the centers of the channels.

Hemocoels employ a similar principle as hemolymph flows over the surfaces of the organs and cavity, giving up and receiving molecules to and from cells on the surfaces. Both microfluidic chips and hemocoels rely on surface to volume ratios for control of extractions.

Porous Monolithic Polymers

Porous monolithic polymers are continuous surfaces of polymer prepared using the channels in the microfluidic chips as molds. Employing a channel and polymer together allows varying the surface areas in contact with the flow. If the polymer fills a large portion of a channel's cross-section, more active surface confronts the moving stream, facilitating more complete extractions from highly dispersed samples. Additionally, using different combinations of porogenic solvents alters the porosities of the polymers thus varying the conditions for the reactions (Ref: Microfluidic Chips).

Scale and Size

Bees and devices are continuous systems shaped for what they do. Because anatomy determines and overlaps with function at so many levels simultaneously, we have few endpoints for measuring rates and distances and correlating the two. How and what we measure depends on how coarse or fine our measuring instruments are. There can be no single true value for a measurement. Surfaces and borders at bee 'sizes' are often self-similar, meaning that small parts are shaped something like the whole. Most natural objects, unlike mathematical objects, only approximate their self-similar forms (statistical similarity). Statistical self-similarity means that something we measure on a piece of a whole object at high resolution is proportional to this same property measured over a larger piece of the object at a lower or coarser resolution.

But measured properties or behaviors are also functions of the sizes of our measuring tools and the scales at which we measure the properties. Size enters our thinking at so many levels. For example, our present theories of continuous systems often start by assuming we use a separate scale for describing the macroscopic variables of a theory and another scale for detecting the microscopic motions of the small particles the theory depends on but that we ignore. For example, hydraulics equations ignore how oxygen binds with hydrogen to form water molecules. The hydraulics equations handle water molecules as point particles. Bulk flows within bees or devices are orders of magnitude greater than the distances between molecules of the hemolymph.

Size, in and of itself, affects almost every aspect of a system's functions. Scaling implies a functional range of sizes. Foundations for scaling relationships lie in geometry. Take any object — a sphere, a cube, a bee, a device. Each possesses the geometric properties of length, area, and volume. Areas are proportional to some measure of length squared, while volumes are proportional to length cubed.

Here's the Rub

Let's see what happens if we change the size of a bee or a device but we keep its shape (or our relative linear proportions) constant. Voila! Something apparently quite new. Let's say we increase length by a factor of two. Areas are proportional to length squared, but the new length is twice the old, so the new area is proportional to the square of twice the old length; the new area is not just twice as large as the old area, but four times as large as the old area. Similarly, volumes are proportional to length cubed, so the new volume is not twice the old volume, but two cubed or eight times the old volume. So as the "size" of a thing changes, all the areas change faster than the linear dimensions, and all the volumes change faster than the areas. These changes in areas and volume have much to do with how bees and devices communicate inside themselves among their parts.

Scaling

Scale is always difficult to visualize and even harder to think about. Scientists and engineers differ in how they approach scale. A scientist might want to understand how the hemocoel works; an engineer might want to build a replica of a hemocoel to separate molecules of the engineer's choosing. The engineer often starts with a theoretical model but immediately must confront questions of scale. The scientist superimposes units of measure on a fluid-filled, constantly changing system where there are few if any stable end points for measurements.

Shape Implies Forces

Our understanding of shape leads us to imagine the forces we think gave rise to the shape. Perfect Euclidian spheres and films form under almost equilibrium conditions (consider a spherical raindrop), but only in the simplest scenarios may equilibrium thermodynamics and statistical mechanics explain all that we see. Let's observe drops of water on a hairless smooth surface of a bee's cuticle under a dissecting microscope.

Condensation and the 'C' Word

Using simple equilibrium physics we can *explain* the spherical cap form of each separate drop, but when several drops coalesce, we immediately have non-equilibrium kinetics. The terrible 'C' word has entered.

Seen at small scales of length the fractal structure of our drops requires a larger intervening surface to hold each fractal drop separate from the surrounding space. But as each drop reaches its critical point, surface tension or the interfacial energy of each drop approaches zero, and a very large wetted surface of cuticle may develop as the drops coalesce. Now our simplistic thinking about a surface separating two drops or phases loses its meaning. Ultimately what is dissolved in the drops and how each drop "unit" combines with its neighbor might allow large surfaces to exist without collapsing or dissolving. Our attention now moves to the junctions between what are the subunits and how their edges contact the water.

It's All in the Edge

Remember the game of taking a photograph of your family and Xeroxing it, and then Xeroxing each successive Xerox image to eliminate the grays and replacing these with lines indicating the edges where the gray areas once joined? The amazing thing is that after many copies you still recognize grandmother. Her face, now just lines, lives concentrated along her edges. Understand where the edges are and where transitions happen, and much information about a system becomes recognizable.

Or take a more relevant scenario: within a hemocoel, polymers of chitin and liquid froths containing protein in a "mixed" system might momentarily be at equilibrium but still beset by forces tending to drive the system away from equilibrium. As a consequence, shapes continually change and grow more complicated. Were we to take our sequence of Xerox snapshots of these changes, we might watch the system progress from a uniform liquid through regular or periodic patterns to a disordered more chaotic form. Add to all this the complications in the transitions brought about as laminar flows become turbulent together with the chaotic behavior of some strongly driven chemical reaction, and we rapidly can assume we can know very little about the mechanisms operating. Were just one of these constraints not present, a very different structure would result. You have the 'C' word again; take your pick: chaos or complexity.

Microns or Nanometers?

So against such physical complexity we primates now must superimpose our measurements. The micron scale is volumetrically ten to the ninth times larger than the nanometer scale, and current technologies provide no mechanisms to control with precision molecules interacting against and within the interior surfaces inside such a complex multidimensional structure as fluid in a hemocoel. Remember a human hair is about eighty thousand nanometers thick, so we use the nanometer scale to measure molecules, strands of DNA and the microscopic structures that determine how steel and plastics perform. In our thinking and in our models, we often ignore the structures of materials at their most fundamental levels unless the energies of these materials in a device or organism exceed the characteristic excitations for the materials. We must remain aware, however, that devices and bees always exploit the ambient physics, and more important, the smaller you are as a device or insect, the more important microphysics becomes for your stability and well-being.

Insect Units

Insects exist at a size larger than nanometers, so what units shall we use? We will measure most often in microns or micrometers. Only sometimes will we use nanometers. The micron unit is easy to keep in mind.

Micrometers Visualized

The diameter of a single human red blood corpuscle is thirty micrometers across. Insects are so large in comparison that they are not like atomically precise devices of nanometer dimensions where the components are discrete molecular parts rather than continuous materials. So insects are not nano-machines that we can model using artificial molecular machines put together by molecular manufacturing.

Nanometer Analogy

Lets see how small really small is. Nanotechnology is based on the nanometer, and a nanometer is one billionth of a meter. We have difficulty making analogies of nanometers with the scales of length we intuitively understand from our everyday sensory world, because nanometers are so extremely small. Michael Mehrle has given us one visual analogy (Michael Mehrle: *michael@w3media.com*). Stretch an imaginary rubber band about a yard long from Los Angeles to New York. That's about 4000 km. Now one nanometer of the rubber band will have been stretched out to 4 mm or 0.16 inches. So whenever we talk in nanometers we are talking about molecules and the distances between them. So most of what we shall consider to happen in hemocoels will take place on the micrometer scale, but still we need to keep the scale of nanometers in mind.

Scale Difficulties

We shall continually encounter difficulties with scale: often we know well the behavior of transport mechanisms on the micro-scale (the scale of individual grains or particles), but we understand transport processes at averaged or macroscopic scales less well. Think of turbulence. One key goal for our thinking is to be able to pass seamlessly from one scale to the next, deriving from the laws governing what happens at the micro-scale those governing the underlying physics of what happens at the underlying macro-scales and *vice versa*. Once we can propose model equations, our next challenge might be to compute the fluid dynamics for multi-phased flows in various situations.

Complexity Magnified

For a moment imagine all the complexity through all the different levels simultaneously. For this part just think of a bee as being a smaller you. The big over arching relationships are similar. You contain approximately sixty trillion (60×10^{12}) cells. At each instant, in each of your sixty trillion cells parts of your genome, a ribbon of two billion characters of DNA (remember in **each** of your cells) produces your proteins. Your genome contains most of your DNA and your genetic inheritance, and, at the same time, the instructions for both constructing and operating all your developmental stages from when you started as a fertilized egg to your death.

Your sixty trillion genomes parallel process the proteins of your life in as many cells continuously all your life. Presumably you make very few errors in reading and decoding your genome, and when you do, you successfully detect and repair these. Your DNA makes RNA makes protein sequence is remarkable for its complexity, its precision and its universality. Of great interest now, your sequence relies on discrete information encoded within the geometry of your DNA. By adopting features of life's cellular and genomic organization, and by incorporating similar or analogous processes within two-dimensional integrated circuits, properties unique to your specific life, (self-replication and self-repair) might in some future way be transferred to our engineered objects. But we are far away still. What are our devices like now?

Divisive Devices: Our Smallest Parts Still Stick

Our best-engineered micro-fabricated micro-machines are mechanical oscillators and moveable mirrors. However, friction and other surface effects as well as clumsy power sources plague even our best machines. Small parts still stick. In contrast, bee systems operate in water within narrow ranges of temperature and pressure, carry built in energy supplies, and utilize the same surface physics more advantageously. To build machines by *picking and placing* atoms in sequence requires forceps smaller than the atoms. Because atoms stick together, we need high energies to pull atoms from one place and deposit them at another, and we must remove any energy we release by cooling. Water batters small groups of atoms in biological systems, so any steering maneuvers we might employ are difficult.

Most devices are products of chip technology, and chip manufacturing may someday evolve to become more biological. Apart from evolving silicon chip technology, we can design chemical computers. Chemical machines process information by making and breaking chemical bonds. Similar technology may create DNA machines closer to how bees do it. A chemical computer breaks a bond, retrieves information encoded in it, and then stores the information in new bond sequences of the machine's making. We may soon represent information in quantum states, such as the spin of an electron. Ever so slowly we are learning to control quantum states by varying electromagnetic fields facilitating advent of quantum computers.

Corralling Molecules

Richard Feynman foresaw nanotechnology and miniaturization of devices towards the molecular level. One use for small devices is design and manufacture of nano-computers and nano-assemblers and combinations of these into large-scale "intelligent" machines having nano-computers as "brains," but problems abound at these scales. How do we control large numbers of extremely small mechanical parts and overcome manufacturing difficulties? Long ago bees evolved to control small moving and non-moving parts as well as their ways of manufacturing. We solved ours with nanotech.

Nanotech?

Nanotechnology is a catchall label for products and processes operating on or very near the molecular scale exploiting systems and materials whose dimensions may be measured in molecules. Up to ten years ago, nanotech considered the average characteristics of blocks of atoms and molecules manipulated only in bulks smaller than grains of sand. Today the nano-world is a nano-cosmos of atoms and molecules ranging from 0.1 to 100 nanometers. Not all workers are studying individual molecules rather than bulk properties, and nanotechnology's semiconductor and biotechnology sectors are growing by leaps and bounds. For example, in 2004 the Federal Government estimated that nanotechnology will have a one trillion dollar economic impact by 2015 (New York Times March 15, 2004 'Bashful versus Brash in the New Field of Nanotech' by B. J. Feder).

Limits? What Limits?

In the 1960s, Gordon Moore realized the transistors on a chip doubled every one and a half years showing exponential growth. This relationship, not a law of nature, is Moore's Law. His intuitive projection of what exceptional engineering might accomplish applied over many decades predicted VLSI or very large integrated circuits. Keynes (1987) states however that wires thinner than an atom and memories less than one electron are not possible and that not only physics, but the astronomical costs, impose intractable limits. Even though today's gate speeds and bit densities are no longer the bottlenecks they once were, microfabrication is still the challenge (Ref: Limits).

Chip Realities

Conceptually, making a chip is like designing and printing an etching on silicon. Using lithography we project a drawn pattern of the connectivity of the components onto a wafer of crystallized silicon. The diffraction limit and need for miniaturization pushed these projections to use the shortest wavelengths possible, but these are still well above atomic sizes. For resolving atom by atom, STM's or scanning tunneling microscopes may follow rows of atoms piezoelectrically to delimit atomic topographies, albeit slowly, but the huge cost of depositing, exposing, etching, implanting, doping, dicing and testing chips during fabrication are still limiting factors. Also as chips increase in size, and the size of each component on a chip shrinks, a single dust particle wrecks greater and greater havoc (Ref: Physics of Devices).

Bottom Up or Top Down?

Many of our smallest machines are downsized versions of our larger ones. We can either build a machine up from atoms and molecules piece by piece, or we can construct a machine top down using such methods as rapidly solidifying jets of droplets of liquid polymer sprayed on surfaces to build up sheets of nano-fibers. A related technology, electro-spinning, forms fibers having nano-scale diameters, but we use tens of thousands of volts to overcome the surface tension that holds the liquid drops together. The voltage pulls charges from the inner surfaces of the drops onto their surfaces crowding surface molecules of the drop causing them to repel each other. As the surface molecules separate, new ones enter the surface from below to destabilize the drop. In stark contrast, epithelial cells of bees secrete layers of chitin fibers and protein bottom up to form exoskeletons using less energy (Ref: Chitin).

Shrinking Big to Small Doesn't Work

As in our printer analogy in Chapter 1, if we shrink conventional micro-electrical motion devices smaller and smaller augmenting these with integrated circuits, we still have not made many working devices, because as devices shrink new manufacturing difficulties emerge. Most technology for our constructing 'small' arose from complementary metal oxide semiconductor manufacturing techniques for making silicon chips. This technology has helped, but we still lack sufficient understanding of synergy theory to unite our fragmentary knowledge of sensing, signal processing, actuating and controlling such minute devices.

What is MEMS?

A MEMS or micro-electro-mechanical-system is a micron-sized device that senses its environment or transports gas, light or liquid. MEMS are micro-machined using integrated circuit technology so MEMS are themselves a chip-level technology. Making one at a time is not feasible, so we make MEMS in batches at low cost. Most with few exceptions are from silicon, contain integrated circuits, and their moving parts monitor motion.

MEMS are sensors, activators or combinations of these. Sensors measure such variables as pressure, temperature, acceleration, flow or chemicals without modifying what they measure. Actuators control fluids or light. Examples are in car airbags, blood pressure monitors, and ink jet printers. Newer MEMS sense vibrations, recognize fingerprints, provide readout displays, store data, switch electrical or optical signals in DVD's, disc drives, phones and micro-mirrors in televisions. MEMS microgyros compensate for trembling hands on cameras and camcorders.

NEMS

NEMS, smaller than MEMS, are nano-scaled systems having dimensions of 10^{-10} m (molecular size) that range upwards to 10^{-7} m or 0.1 to 100 nanometers. Studying nano-sized structures involves understanding the physics of building up molecules from atoms. A nano-sized process of all insects involves building up chitin, the major material of the exoskeleton, from chitin's components. NEMS employ novel materials such as carbon nano-tubes, quantum wires and quantum dots. Our 'nanotech revolution' involves problems of mass-producing nano-transistors and nano-diodes, nano-switches and nano-logic gates in order to construct nano-scale computers having terascale capabilities. Can we do it? Probably; but perhaps with a few changes in our thinking, insects might help.

Generalities

Mechanical systems lack the similar modularity and topological simplicity of electronic circuits, so small mechanical devices are hard to manufacture. MEMS devices compared with larger machines, have fewer rigidly linked parts, and more are intrinsically compliant. As with chips, we can use planar lithographic processes to make masks for sequences in constructing MEMS.

Generally then, both NEMS and MEMS are micro-assemblies of parts having electronic and mechanical functions with NEMS being much smaller than MEMS. We house and integrate these latter assemblies on a single silicon chip.

A Generic MEMS System

To appreciate just how far we are from building bees, a current generic MEMS system might contain a micro-pump, a flow sensor, and an electronic control circuit. The pump delivers a variable flow into a micro-channel, and the circuit controls the rate of pumping. How do we build it?

First Simulate

To simulate a MEMS system on a computer, we first model each component of the system. Then we make a coupled liquid simulation, but disparities arise both at temporal and physical scales. Unlike an integrated circuit for which we have many programs that can test for errors in design, MEMS lack such verification tools. Having strong linkages between domains of energy in a device makes analysis difficult as computer models rapidly grow unwieldy and computationally prohibitive. So we simplify yet again.

Then Again and Again

We can simplify a model by mathematically lumping its functions together, by reducing its resolution or graininess, or by lowering a model's dimensions. We do almost anything to make the simulation work better; but even complicated mathematical descriptions quickly grow too simplistic, especially if the model is to describe a specific geometric domain or transitional region where our understanding of the underlying meso-physics is incomplete. Remember, bees live at dimensions where our knowledge of the physics a bee encounters daily is largely unknown. Again, one reason for our incomplete understanding is our reliance on overlapping scales in our descriptions.

The complex forms of macro-sized systems and MEMS devices may be incomparable. Macro-sized mechanical devices function in multiple energy domains and utilize many components. Some share topological boundaries, such as fluids bounded by moving parts. Design and manufacture of the more complicated macro-mechanical devices employ several techniques. We have difficulty integrating diverse techniques and tools into generalized sequences or programs because often no general relations or even correlations between a device's form and what it does exist. Example: an automobile production line.

VLSI Devices Obey Simple Laws

Very large-scale integration or VLSI is the current range of size for miniaturizing microchips. VLSI refers to microchips having hundreds of thousands of transistors on each chip. We can use structured design methods to build VLSI systems, because a computer chip functions only within a single domain of energy, and chips obey strict rules. Each of the many transistors on a chip maps directly, one to one, from function to topology, and each transistor obeys simple rules for interconnections. An example of such a 'rule' is Kirkoff's current law that states that the current entering a node equals the current leaving it. There is an analogous law for voltage. Such simple laws used to model circuits have far reaching consequences, as these rules are the starting points for analyzing any circuit. Unfortunately, we still must find similar basic current rules for flows of energy and information for both MEMS and biological systems.

Bulk Controls Electrons

Devices are bulkier than bees. How devices employ their bulk depends upon how they control the diffusivity of electrons. Here, follow two examples in which we compare a device and its counterpart in insects.

Diaphragm Pump Compares Unfavorably With a Bee's Heart

One diaphragm pump, utilizes a piezo-electrical actuated microdiaphragm to dose liquids and gases. Its chip measures $7 \times 7 \times 1\,mm^3$, and the diaphragm displaces 2 mm a minute. The pump tolerates bubbles and primes itself. (Remember, when you are so small, bubbles always give you troubles.) The pump ejects liquid against a 500 mbar head of pressure. Mechanical parts include a valve unit, two passive check valves, and a disc actuator. The latter periodically deflects a pump diaphragm. Parts are of silicon. One adjusts the rate of pumping by setting a frequency or amplitude through a driver module. Review now the bee's heart in Chapter 1 and compare.

Acceleration Sensor Compares Unfavorably With a Bee's Mechanoreceptors

Mechanoreceptors detect distortions of the bee's body by monitoring the bending of cuticle. Distortions occur from brushing against objects, the bending of appendages during walking and flying, and from vibrations transmitted through the air. Mechanoreceptors contain hairs that attach to nerve cells. When the hairs bend, action potentials from the nerve cells inform the nervous system about positions of body parts and rates of change of these bending movements. Bees and other insects also possess campaniform sensilla. These organs are found in areas of their body surface that are subject to stress. Sensilla are projections of thin cuticle raised into thin domes about five to thirty micrometers across. Deformation of the domes activates neurons.

In contrast, one manufactured acceleration sensor is a chip and sensor electronics contained within a DIL40 silicon housing. To build this sensor, one embeds oxygen atoms in silicon and then 'dopes' a monocrystalline epitaxial layer several mm thick on the silicon. A cap sensor in this surface having its 'sensitive axis' parallel to the surface of the chip is held near electrodes suspended on small silicon beams. Moving or accelerating the device increases the capacitance between a fixed and a moveable electrode, so capacitance increases on one side of the device and decreases on the other side. Each capacitor reads out separately to give a differential measurement. Circuits on a second chip sense the differential capacity and convert it to a voltage that is proportional to the acceleration. Acceleration sensors are used now in auto crash detection, controlling vehicular dynamics and for measuring shock.

Example of a Sequenced Array

We may employ a series of devices linking energetic domains: mechanical, electrical, fluidic and optical. Micro-mirror devices link together several energetic domains, but not as well as bees link these. One device entails suspending a plate from two beams and then allowing two parallel-plate actuators to deflect the plate electrostatically. We may also orient individual micro-mirror devices into arrays within larger systems. As in biological systems, the robustness of any pattern of connections lies in how well each component coordinates with all the others. As we shall discover, the bee's circulation connects and coordinates all parts of the bee machine to exercise continuous local control over a maximal number of metabolic interactions.

Miniature Parts

We can also miniaturize conventional electronic computing devices using molecular transistors and quantum dots. Quantum dots are nano-sized single electron transistors, each electron, because it is a single charge, can "store" information. Also called q-dots, dots can be of silicon and just a few atoms across. One advantage is that q-dots glow brightly in ultraviolet light producing a different hue depending on a dot's size. Two nanometer-sized dots of a substance may glow green but five nanometer dots glow red. If we coat a dot with a material that makes the dot adhere to specific molecules, dots may be injected and followed through a circulation to see where they adhere to their targets. In this way, we might watch flows and processes.

We can also use quantum dots as nano-switches to process information. The downside is that nano-switches alone have no memory capabilities. And now comes the even harder part. To build really small we must also design and build logic gates and registers on the scale of single molecules. In even our best micro-scenarios, quantum dots contain a discrete number of electrons that move around as a superimposed electromagnetic field varies. How do bees operate without electronic components? They employ the energy of sunlight captured in the chemical bonds of carbohydrate molecules.

Energy

To supply the power to perform work, bees, unlike devices, ultimately employ sunlight's energy trapped in the carbohydrates of green plants. Energy enters a bee directly as light and heat from the sun or as chemical bonds stored in honey or nectar. Metabolism occurs in discrete steps that break down or catabolize the molecules of carbohydrate, each step releasing small quantities of energy, through a sequence of enzymes. One sequence, the aerobic pathway, works when oxygen is immediately available, and another, the anaerobic pathway, functions when oxygen is less available. Energy from either of these paths transfers to an intermediate molecule, ATP or NAD, through the process of phosphorylation. ATP and NAD are coins in the cellular economy. These molecules can ferry energy to places in the cell where synthetic or energy-requiring reactions take place, allowing cells to build their own specific compounds on location.

Foreshadowing: Emergent Levels of Metabolism

Units of metabolic function are highly variable. 'Units' can range from a single enzyme molecule, to cellular organelles to cells to organs. Interconnections within and between these levels of metabolic function appear seamless. Units of function increase in size from molecules, through tissues, cells, organs and organ systems to a bee or to a population of bees in a colony producing heat warming itself in winter. Each level has its own terminology for description including units of measure. As the size of an individual functional unit increases, at each level and between levels, new emergent functions arise. Topologically, the visible anatomy of a bee shares many distinct boundaries. Examples include cell membranes and the layers of epithelial cells that line the surfaces of organs, but most boundaries between functions remain indistinct. Unlike networks of electronic components in which charge flows through wires, most topological lines between biological functions are indistinct. Some direct mapping between function and form occurs most readily at macro-sized or organ or organ system levels, (heart as a pump) and at macromolecular levels (DNA to RNA to protein), but most intermediate mappings must still be worked out. Lacking detailed nesting maps, we can build at best very imperfect models.

The Monkey's Problem: Analyze Emergent Functions?

So we are back to the monkey's problem again. With varying degrees of difficulty we can physically break bees and devices down into their subunits. Extracting macromolecules or cell organelles from homogenates is comparatively easy, but fully isolating connected cellular units from the matrix of the central nervous system, for example, is exceedingly difficult. Disassembling an organism or a device at any level of resolution disrupts its emergent functions, so that much emergent information gets lost irretrievably.

Presumably sometime about 2040 or so, parameters for chip devices will finally have attained their fundamental physical limits.

Comparisons: Bees

See Table 2.1 comparing bees with our devices. Bees overlap levels of function. Cells coordinate into tissues and tissues into organs. Each component or level of organization excels at doing different work. Components self repair and reproduce themselves in

Table 2.1 Organization of Bees and Devices.

	Bees	Macro-machine	Very Large Scale Integration	Micromachines
Energy	Sunlight metabolism	Multiple, coupled domains	Single domain	Multiple, coupled domains
Units of Organization	Molecules, cells, tissues, organs, organ systems…	Wide range of non-modular, and hierarchical units	Small set of primitive subunits	Intermediate range of primative units
Dissection	Difficult to easy	Difficult	Easy	Limited
Topology	Mostly indistinguishable boundaries	Shared boundaries indistinguishable	Distinguishable boundaries	Shared boundaries indistinguishable
Mapping Between Function and Form	Highly variable	No direct mapping	Direct mapping	Difficult but some direct mapping
Interconnection Rules	Very complex	Complex	Simple voltage/current laws	Complex
Geometry	Integrated shape, function, and kinematics	Shape tied to function and kinematics	Not a problem as planar	Monolithic and compliant now
Manufacture	Sexual reproduction	Machining in 3D over wide range	Planar lithography	Planar lithography

concert with the system as a whole. Bees work at everyday temperatures and pressures and are masterful conservers of materials and thereby avoid unnecessary redundancy and excess weight. Bees adjust to change and reproduce themselves.

Devices

Devices constrain energy to perform specific functions well but do not adapt well to superimposed changes. Exceptions include self-sealing tires. To make devices and for them to work often requires exotic materials and extreme energies and pressures. Devices do not increase their own diversity. However, our devices can approach organisms in their size and efficiency once we learn how to make them more biological, perhaps even enough biological to resemble insects.

Manufacture

We must develop user-friendly tools for computerized analysis and simulation to design and manufacture more complicated MEMS. Because to function MEMS must couple between domains of energy, we can simulate and verify a design before building it, but how a device actually performs under real conditions is impossible to simulate completely. Unlike integrated circuits, where we may incorporate algorithms for the design of components into their manufacturing programs that we join with programs to automatically check for errors, current manufacture of MEMS is harder to check.

Comparing MEMS to larger machines helps us understand why. Compared with macro-mechanical devices, MEMS have far fewer rigidly coupled mechanical components. Instead, MEMS components frequently possess an intrinsic springiness or compliance. Manufacturing compliant components allows us to employ structured designs and planar lithographic masking to make compliant parts systematically for predetermined specified transmissions of controlled motion and force. In this way, MEMS devices share many features with VLSI devices. Hence, using a mask design in a manufacturing sequence may aid sequencing designs to simplify the processes.

Chip Manufacture: A Dirty Problem

Lithography lets us see the features of a device we intend to build. However, the diffraction limit has pushed our optical systems towards shorter and shorter wavelengths that still exceed atomic dimensions. Using atomic force microscopes and scanning tunneling microscopes we can even scan a sharp tip over a sample to measure piezoelectrically the deflections of a cantilever or the electron tunneling current to measure the atomic topography.

If we limit manufacture of devices to planar processes, we vastly limit our ranges of motion for our deformable structures. If we can limit motion, a device's shape and behavior become less complex than those of macro-mechanical systems, and for less complex designs we may use photolithography. But, even here we get stuck. As features shrink, dust and other particles get magnified. As with bubbles in small tubes, at chip sizes a very small defect may disable an entire part.

To circumvent the need for ultra-cleanliness, we can design MEMS expecting that most of those we assemble will be faulty. But even so, we can still rewire hierarchies of components into complex modules after we test the components separately and together.

Looking closely at our MEMS devices shows us that devices have progressed from jointed rigid mechanisms towards compliant and deformable mechanisms, so in these respects, our machines may be more like bees. Now it should soon be possible to use functional specifications alone to create multiple systematic designs that range over rigid and compliant structures to support loads and also to control the transmission of force and motion. After we determine components and the displacements of parts within a device, we may then be able to make masks for lithographic reproduction almost automatically.

Manufacture Close on the Bee's Knees

Producing nano-machines may someday be more bee-like because even today by using sophisticated polymerization techniques, we can build up large linear molecules that then spontaneously fold themselves into three-dimensional structures without need of brick-by-brick pick-and-place construction. We may even encode directions within a molecular array for sequencing additional sub-units as well as their eventual functional complex in the contained chemical sequences. Such would be a useful direction for our research, because as every bee knows at a gut level, linear sequencing is unbeatable in its efficiency. We are ever closer on the bees' knees. We can now evaporate colloids of nano-particles, each particle smaller than three nanometers in diameter, onto surfaces to form crystallized two- or three-dimensional arrays. Analogous arrays may soon store data.

Showcase Bee Manufacturing: Chitin

Bees construct their chitin cuticles from new and recycled precursors. Bees linearly process recycled and new components through their hemocoels. We shall refer to this model again.

What is Chitin?

A bee's exoskeleton of cuticle is a continuous material around a soft body. Much of insect success is due to the flexible properties of cuticle. Cuticle is a network of polysaccharide fibers (chitin) embedded in a protein matrix. This fibrous composite combines the intrinsic strength of the embedded chitin fibers with a high degree of toughness (the ability to absorb the energy of impact). Strength of cuticle to a great extent arises from the forces existing between molecules of the hard matrix binding these to the surfaces of the fibers. Chitin not only protects; it is essential for growth (Ref: Chitin).

Growth

Growth in insects occurs between molts, and in bees molting occurs as the pupa turns into an adult bee. Before a developing bee sheds its exoskeleton, the larva, a little white worm growing within a waxen cell of honeycomb being used as a nursery, enlarges itself and acquires the adult form. As much internal reorganization of the pupa's body takes place, chitin and protein subunits from the old exoskeleton recycle within the hemocoel, so that many of these same components find positions in the new adult cuticle that forms underneath the old. Producing normal chitin requires a precise spatio-temporal ordering of molecules. This ordering depends on the geometries of receptors and their positions in space.

Sorting and processing molecules in the hemocoel begins in the hemolymph. Small molecules in one part of a bee end up in complex structures in another. Water is a natural starting point for the insect's molecular manufacturing. Hemolymph, largely water, is dense, highly disordered, and simultaneously transports a diversity of cells and molecules. Manufacturing cuticle includes: acquiring and ordering molecules from the hemolymph, transforming streams of incoming molecules into streams of product, and then storing or arranging these strings to build the complex three-dimensional object that is cuticle. To analyze these sequences, we must ultimately understand: the assembly sequence, how molecules are transported, the timing of synthetic cycles, their energy requirements, the error rates and the error sensitivities.

Contained Energy Supplies?

Few devices have these. So far no micro-engineered system comes close. One reason is that batteries are the contained energy sources for many machines. Batteries are heavy and come in lumps. Worse, batteries, as we shrink them, deliver less and less power per unit weight. Shrinking batteries brings us back again to our problems of scale that evolution has solved.

Biological evolution guarantees economy of time, materials and motion and a maximal return for metabolic outlay. A bee, like a machine, performs a task, is constructed in a process, uses power and operates using built in or supplied information. However, bee power is not battery power. Batteries have the extreme disadvantage that battery function as a function of mass falls rapidly as a device shrinks. We might think of a bee as being more like a self-winding watch that siphons energy from the environment as energy is needed. Nonetheless, bees have solved the nano-difficulties of coupling between electrical, mechanical, thermal, radiant, chemical, magnetic, acoustic and fluid domains of energy (Ref: Self-reproducing Machines).

Webs of Levels

On the other hand, our micro-electrical-mechanical-systems and their smaller brethren are heavily biased away from bees as both evolved from micro-technologies that depend largely upon electronics and silicon. Devices lack biomechanical foundations. MEMS, however, to their credit, are often hierarchical systems composed of layers, and we may think of bees as being composed of telescoping layers too. Both devices and bees are emergent systems starting with molecules passing through larger and larger webs of parts to the whole. At each level, behaviors are greater than the sum of the parts at that level. Each bee is built up of complex subsystems governed by physical, chemical (hormones) and electrical (neural) controllers for each component.

But compared with a bee's enmeshed levels of organization that are so much more complete and intricate, our NEMS and MEMS devices are incredibly simplistic.

Neural Coordination

Conceptually, many of today's micro-devices are not much more than miniaturized versions of our larger devices. We need devices that, like insects, operate in size ranges extending from simple diatomic molecules up to biological molecules having 10^6 to 10^9 atoms. Larger biological molecules such as proteins and nucleic acids usually fold up into complex tertiary structures to occupy three dimensions. Many biological molecules may transform themselves in time as well, so even at this elementary structural level, biology has the edge over silicon. Still let's not forget: much coordinated behavior and motion as well as vision and sensing, decision making, memory and learning in bees results from neurons transmitting electrical information over neural networks within milliseconds. Transmission of neural coordinating information is complex. Unequal separations of ions: sodium, potassium and chloride, across cell membranes create membrane potentials, and changes in permeability propagate over distances using a unique spike or action-potential mechanism as well as slowly varying membrane potentials. Nonetheless like devices, insects employ inorganic components for control and coordination.

Copper, Iron and Ordered Structure

Biological molecules such as proteins often coupled to inorganic atoms such as iron or copper may control molecular interactions and biochemical pathways. Most solids do not usually consist of a single phase, because defects and boundaries exist across different kinds of domains. It is what happens at the boundaries, often between domains of energy, that determines how information and materials pass through the system. Engineering utilizes the macroscopic as well as the microscopic, and micro-technologies let us work within larger and smaller scales at the same time. Macroscopic materials having a crystalline structure (a complete long range ordering of molecules) and liquids and glasses having short-range order and little long-range order, as well as gases having little short-range order, might all be coerced into operating together within living systems.

When You Are Small Your Rules Are Different

Insect miniaturized systems are incredibly robust and successful. Different rules dictate success. The small sizes of many insects that take different forms throughout an insect's life: egg, larva, pupa, adult let insects adapt to ecosystems worldwide, where they form major components of many diverse ecosystems. Let's learn the new rules.

Bees as MEMS: A Summary

Bees control and manipulate fluids within their bodies using complexly interconnected systems of pumps, valves, manifolds, tubes, connectors, reservoirs and transducers. Bees are best at interfacing; much better than we are! Bees have wonderful edges. Biological systems masterfully coordinate fluid mechanics near and across complicated interfaces to adjust minute volumes of fluid in parallel within different compartments while at the same time, but only when necessary, blending them while also permitting volumes of fluids to adjust within several body compartments simultaneously. A bee's fluids contain micro-structural elements, and these elements interact via colloidal, hydrodynamic, and Brownian forces. If one's body is as small as a bee's, one desiccates in the sun and over-hydrates in the rain, all the while adjusting water balance while performing the daily chemistries of living. All are major obstacles. Bees also function well within the physiological extremes of temperature and pressure. Devices on the other hand, that employ moving fluids at microscopic dimensions, range from endoscopes to labs on chips. Devices transport minute quantities of liquids or gas through networks of microchannels. Controlling micro-flows in our devices includes: pumping, electro-osmotic flows, electro-wetting and thermo-capillary pumping. For small devices to employ such technologies requires that we not only micro-machine their interior channels but employ high temperatures or kilovolts to drive flows. We must understand a device's microphysics and all its contained flows. Bees do better than we do without all this knowing. Can we learn our microphysics from them? Once learned, can we expand chip technology to include this type of biology?

> 'Our imagination is stretched to the utmost, not, as in fiction, to imagine things which are not really there, but just to comprehend the things which are there' (Richard Feynman).

Chapter three

BEAUTY BEFORE THE BEAST

A. GRAPHS

Graphs Before Models

We introduce elementary graph theory before we show that mathematical graphs reveal important differences between closed and open circulatory systems. We then introduce modeling. Basically, one way to compare distribution systems, be they natural or devices, is to compare their graphs and then build models. Graphs come before models, because before we build conceptual models, we should learn the geography of a system's connectivity. What parts connect with what parts? How far apart are they? Graphs abstract away the trees and bushes covering complex landscapes revealing the contours of the hills. A graph in its simplest form and sufficient for our context is a set of points connected to each other in some way through a set of lines.

Graphs in General

A generalized graph is a set of vertices and edges. We usually represent vertices as dots and the edges as lines connecting the dots. The edges, the lines, also called links, connect the vertices. A more formal definition is a graph, G, is a non-empty set of elements, the vertices, and a list of unordered pairs of these elements, the edges. The set of vertices of graph G is the vertex set of G; we denote this set by $V(G)$, and the list of edges is the edge set of

G, denoted $E(G)$. If v and w are vertices of G, then the edge vw connects or joins v and w.

Why graphs are so useful is that graph theory treats only the number of elements in a network and their relationships to each other, both in terms of the characteristics of the edge set. First some terms.

Order and Size

A graph's order or n, is how many vertices it has given by $V(G)$. M, the number of edges in the graph $E(G)$ determines a graph's size. Graphs can represent all kinds of networks. Vertices may be organs: heart, kidneys or brain, or chips, valves, or some anatomical or functional combinations of these. Edges represent predicted or defined relationships between the vertices. Edges might be roads, flight paths, wires, blood vessels, nerves, tubes, lightpaths, routes of diffusing molecules through the hemolymph, or combinations. The elements of a graph and their connections can represent any groupings of characters we choose. This one generalization makes graphs such powerful tools. To make a graph we abstract everything else away. As with the Xeroxed picture of grandmother, what is remarkable is how much information remains in her connected edges.

The World Wide Web: A Graph

Consider the World Wide Web as a graph. The Web's architecture as is a hemocoel's is non-engineered. The edges are a virtual network of hyperlinks, connecting more than eight billion web pages. These pages, created by tens of millions of independent people at separate locations are the vertices. Large computers can scarcely contain these data points, but after cataloging what we know of each vertex and edge, we can create a model simplifying the WWW but retaining some properties of its larger graph.

From smaller representative data sets we better appreciate global and local relationships, and later we apply these to the larger graph. For example, growth at vertices both in the web and in hemocoels, is decentralized but self-organized. The web contains a large, strongly connected core of prominent sites through which every page can reach every other page over a shortest path of sixteen to twenty hyperlinks. Distances through this core are much shorter than across the whole web. Local actions grow new vertices near the old ones, and the chance that an existing vertex or link receives a new link is proportional to the number of links the vertex already has, resulting in a power law distribution of links (Ref: Power Laws).

But growth patterns differ: Classical random graphs are regular graphs (almost all their vertices have the same expected number of edges), but in real-world graphs a few vertices have very large numbers of edges. We can extend our random graph model to include these new forms (Ref: Algorithm Design).

Understanding local behavior at neighboring vertices of the WWW lets us reason that an unusually large number of links among a small set of vertices, or web pages suggests these pages may be related to the same topic and that the group may even be a signature for the area. For example, in the hemocoel, the edges leading to vertices at or near the openings of the Malpighian tubules might suggest excretory metabolites and processes, so computing these network flows might partition the hemocoel into regional functional graphs. To build is to understand.

Properties of Graphs

Graphs may be unrestricted, simple, sparse or connected. In unrestricted graphs, the edges have no inherent direction to imply a symmetric relationship between the vertices each edge connects. We may leave edges unweighted and not give these strengths *a priori.* An unweighted edge is important only in its relationships to other edges. In a simple graph, we forbid multiple edges that connect the same vertices. A forbidden edge would be one that might circle around and then connects a vertex with itself. A graph may be sparse. For an undirected graph the maximal size of $E(G)$ equals $(n > 2) = n(n-1)/2$, corresponding to a fully connected or complete graph. Sparseness implies M, the graph's size, $\ll n(n-1)/2$. Lastly, we may create a connected graph. In a connected graph, one may move from any vertex to any other vertex by traversing a path of only a finite number of edges. A circulation, be it open or closed, is a connected graph, because a drop of blood from one region may circulate to any other.

Restricted Representations

Our choices and assumptions restrict realistic graphical representations. Graphs, by themselves, introduce a minimum arbitrary structure into our analysis, but even so, graphs are one useful basis for modeling and comparing complicated relationships in organisms and devices. We make directed edges or one-way streets (arterial blood flow leaving or venous blood returning to the heart for example), and when we create directed edges, some relationships automatically become more important than others. Many real networks are unconnected, but multiple relationships exist simultaneously between the same set of elements, as with husbands, wives and lovers. For each relationship, its graph would be different but the vertices are the same.

What Graphs Show

Graph theory helps explain practical operational problems. Example: The shortest path problem. Imagine a map of a closed circulation supplying organs, cells, or regions. As in a road map or routing diagram the vertices are the organs or points of service, and the edges are the connectors linking the points. Our map is a connected graph with a non-negative number representing each edge. As in a roadmap we may choose shorter and longer sequences of edges. What is the shortest path? Edges need not be distances but may be times or costs of travel or measurable metabolic or energetic costs of flow. We can construct an algorithm for the graph. We find an upper bound by taking the longest path and calculating its length. Our map is now a weighted graph, and we must find the route having minimum total weight. If each edge has unitary value, the problem reduces to finding the shortest path as being the fewest edges linking our two points. A variant of this problem is the postman's problem that circulations have solved during evolution.

Postman Problem

Here a carrier wants to travel the shortest distance through a neighborhood and return to his truck. Using a weighted graph for this network, the weight of each edge being its length, how can he 'walk the walk' over each edge but once? Or, more biologically speaking, how might carrier molecules such as hemoglobin cover the shortest distance through the circulation delivering oxygen to all points? Here we have many carriers in concert, but all should travel the least distance and avoid traversing too many paths more than once thus wasting heartbeat energy. Our weighted graph now corresponds to a circulatory network. For vertebrates evolution operating over millions of years discovered a closed circuit of minimal total weight that includes each edge at least and usually no more than once.

Closed Circulation is a Eulerian Trail

The closed circuit of a vertebrate circulation is a Eulerian trail or circuit. The circuit starts and ends at the same vertex, the heart. In a Eulerian circuit each edge is traversed only once, and blood in this optimal situation visits each component of the circuit only once. Without doing the math, if the graph is Eulerian and the circuit is a closed trail of the Eulerian type, we may solve it using Fleury's algorithm, but if the graph is non-Eulerian we also have an algorithm (Ref: Graph Theory).

Trees

Using a tree graph, we can create a circulation pattern of n organs or points in which a volume of blood can travel from any of the n points to any other point. The tree graph pattern in the bee illustrates blood flow from the heart to the brain through the dorsal vessel and aorta. This graph will be useful when we consider open circulations in which materials do not necessarily pass through the heart each time before going to a new destination.

If we must shrink a closed system for economic or anatomical reasons minimizing tubes and distances, then our graph having the n points as vertices and the connectors as edges forms a tree graph. The problem now is to find an efficient algorithm for deciding which of the $n^{(n-2)}$ possible trees connecting all our points traverses the least distance and, consequently, uses the least materials or energy. Evolution has created closed circulations that do this. Solutions assume we can measure all distances. Again, we can formulate the problem using a weighted algorithm and what's known as the greedy algorithm to find the optimal solutions.

A greedy or single-minded algorithm performs a single step over and over until the steps can no longer be repeated. The algorithm then chooses the next step and continues until stoppage. Results may not be perfect but are a first approximation. We have no algorithm, for example, that generates a pattern of coloring using the fewest colors for every map, but if we color in as many regions as possible with one color before choosing the next color, and then choose a new color only after we have exhausted the last, and we repeat this sequence until all patches are colored, we have followed a greedy algorithm.

At one point in the process a local optimum becomes a global optimum. Similar logic helps find a dominating set or the group of vertices having all other vertices in the graph as its neighbors. One starts with the vertex having the most vertices in the graph linking to it and then chooses the vertices of the next largest degree and so on until we find the dominating set. A hemocoel containing declining volumes of hemolymph passes through states in which local deep pools grow shallower progressing to global behavior as the volume of hemolymph becomes a two-dimensional film wetting all surfaces (Ref: Greedy Algorithm).

Small World Graphs

Many systems display three overlapping characteristics. Their graphs are sparse; their vertices cluster, and the graphs are of small diameter. Graphs having the three properties of sparseness, clustering and small diameter are small world graphs (Ref: Small World Networks).

Sparseness

Consider a graph of the connectivity within an animal at any resolution. Despite the huge numbers of vertices, connections in an animal have relatively few edges and are sparse. In a graph having n vertices, the maximum number of edges is $n(n-1)/2$, or roughly $n^2/2$. (I consider here only "simple" graphs, as opposed to multi-graphs, where more than one edge can join a pair of vertices.) In large graphs of the real-world variety, the number of edges is generally closer to n than to $n^2/2$.

Clustered Vertices

Vertices tend to cluster in that the probability increases that two vertices whose edges link to a third vertex also link to each other. In other words, each vertex has neighbors each chosen using a distribution that weights vertices by their current degree. Thus, on the World Wide Web the edges of the graph of web pages and pages linking to them are not uniformly distributed but form clumps, knots or hubs such as the website for a major newspaper or Google that receive many more than the average number of links to other pages.

Small Diameters

Graphs in the real world tend to have small diameters. The diameter of a graph is the longest or shortest path across it depending on one's definition, or in other words, the length or the number of vertices traversed during the most direct route between its most distant vertices. In a closed circulation, blood follows the same circuit, but in a hemocoel routes may be direct or longer.

Only connected graphs have finite diameters. A connected graph is all in one piece, and it must have at least $n - 1$ edges so its largest possible diameter is $n - 1$. At the opposite extreme, a complete graph, having $n^2/2$ edges, has a diameter of 1, because one can pass from any vertex to any other traversing a single edge. Graphs having edges closer to the minimum than the maximum number may be of large diameter. Clustering could increase the diameter even more, because edges used up to make local clumps leave fewer edges available for connections spanning longer distances as occurs with regulated airline routes. In a hub-and-spoke system, travel to many cities is not over the shortest distances because stopovers involve hub cities. Nevertheless, the diameter of the World Wide Web and other large graphs appears to hover about the logarithm of n, which is much smaller than n itself.

Simplified Models

To help us understand very large small-world graphs, we consider two simplified patterns occurring at the extremes. Both will be useful when we compare closed with open circulations.

Lattice Model

The simplest of all graphs are highly regular lattices in which every vertex joins to just a few neighbors. 'Lattice' brings to mind a two-dimensional square grid, but a lattice graph may have other geometries. A minimal lattice is a one-dimensional structure, like a chorus line with dancers holding hands. Bringing this linear lattice line into a circle and joining its two ends forms a ring lattice or cycle. A nearest-neighbor ring lattice with n vertices has n edges, (each dancer is a vertex) and every vertex has degree 2, meaning that two edges meet at each vertex. When edges extend both to nearest neighbors and to next-nearest neighbors, the ring has $2n$ edges and vertices of degree 4. A ring lattice is a poor example of a small-world graph. It is suitably sparse, having just n edges in the nearest-neighbor case, so in one sense it is highly clustered, because all its edges are 'local.' But the diameter of a ring graph cannot be small. The only way to traverse a ring graph is to pass from neighbor to neighbor; a lattice has too many vertices. It is like taking a streetcar that stops at all the stops along the route. One gets there faster by taxi. The diameter of the nearest-neighbor ring is $n/2$. This is much larger than $\log n$.

Maximum Randomness

A lattice graph is highly ordered, but a random graph is its opposite: maximally random. To create a random graph, begin with n vertices and no edges. Consider **every possible** pairing of vertices. For each pair either draw an edge having probability p or do not draw an edge with a probability of $1 - p$.

We can predict the outcome for our two extreme cases: If $p = 0$, our random graph is edgeless. If $p = 1$, our graph is a simple graph, or more specifically, a clique graph in which all pairs of vertices are adjacent to each other. Between these two extremes, we expect intervening graphs to have about $pn^2/2$ edges. The Hungarian mathematicians, Paul Erdös and Alfred Rényi, studied the number of vertices, edges and their connections: all determined randomly. Most proofs about random graphs consider, "almost every" random graph to have a certain property. "Almost every" means that as a random graph grows larger and larger towards infinity, the probability of our certain property occuring anywhere in the graph approaches 1. So if the probability of having an edge p is greater than a certain threshold probability, then almost every random graph is connected. This statement is important. The Erdös-Rényi random graph can be made to be as dense or as sparse as we want it to be just by adjusting the edge probability p.

Changing Edge Probabilities

The diameter of a random graph tends to be small (in some cases too small). But random graphs do not form clusters because their edges occur independently. Independence of an edge means neighbors of neighboring vertices are no more likely to be linked than any other randomly chosen vertices. Does this begin to sound like a graph of the potential connections within a hemocoel especially one whose volume of hemolymph is changing? Stay tuned.

"Real Live" Graphs Are Mixtures

For the most part, nature's connections are neither regular nor random. A cell or organ communicates mostly with its immediate neighbors as the lattice model implies, but many cells and organs carry on more distant relationships. Cells at one location may produce hormones having global effects over long distances. We form our links on the World Wide Web not at random because we intentionally link to specific pages. So most graphs of "real live things" end up as mixtures somewhere between order and randomness, our two extremes.

Interpolate Between Extremes

Strogatz interpolated and found that his modified graphs described well many diverse situations (Ref: Small World Networks). He began with a regular ring lattice and then 'rewired' some of the edges to introduce randomness. To do this, examine each edge of an original lattice in turn. Either leave the edge as you find it or redirect the edge to another randomly chosen location. Your decision to rewire an edge is governed solely by a probability p, which you adjust over the range from 0 to 1. If p equals 0, you leave the edge alone, and the lattice remains unchanged. If p equals one, you transform your lattice into a random graph.

In analyzing their rewired lattice graphs, Watts and Strogatz did not examine the shortest path between the most distant vertices, the diameter of the graph. They studied the minimum path length L averaged over all pairs of vertices. They found the minimum length of a path changed markedly as rewiring probability increased, and they rewired more and more edges. L is at its maximum in the regular lattice, but L falls steeply after rewiring just a few edges.

Measuring Degree of Relationship: Clustering

How tight or loose are the relationships in a cluster? In a hybrid or mixed graph we define a clustering coefficient, C. To calculate C, list all the neighbors of a vertex, count the edges that link to its neighbors, and then divide the sum by the maximum number of edges that could possibly exist among the neighbors. Now repeat this operation for all the vertices and take the average. In contrast with the path length L, the clustering coefficient C stays large until the rewiring probability becomes rather large. Hence, over a wide range of p values, local connections dominate the graph between nearby nodes, but just a few shortcuts, like freeways connecting densely populated cities, may suffice for efficient long-distance connections.

Shrinking Diameters

We now see that randomly rewired connections can shrink the diameter of a lattice, facilitating long-range connections, but shortcuts are not useful if we cannot find them. Substances diffusing within an open circulation find the short paths each time, but closed circulations have unalterable longer and shorter paths built into their patterns of vessels.

Finding Shortcuts

Begin with a two-dimensional square lattice, a checkerboard, where each vertex links to its four nearest neighbors. Now add long-distance connectors, but do not add these purely at random. For each vertex, assign a rank to all the possible destinations for your shortcut link, and as would a highway planner, base your rankings on how far the shortcut extends from the source vertex.

The probability of choosing a vertex at distance d is proportional to d^{-r}, where r is an additional parameter of your model. If you set r equal to 0, then you select all destinations at all distances with uniform probability, and your model is now just a two-dimensional array of the Watts-Strogatz model. If r is large, now you have the best chances of choosing only nearby destinations, and you have barely altered the original form of the lattice. The crucial value turns out to be $r = 2$, when the probability obeys an inverse-square law.

We can easily traverse graphs where $r = 2$, not because they have the smallest diameters (they don't) but because we have an algorithm for finding a short path through them. The algorithm is our simple 'greedy' one. To find a route from node a to node b, list all the edges emanating from a, and then choose the edge that takes you closest to b, measure lattice distance (you do this when reading a road map). Now repeat the same map reading sequence beginning again from this intermediate point, and continue sequencing from point to point until you reach your destination. The greedy algorithm being most efficient when $r = 2$ determines the spectrum of edge lengths. No other algorithm performs better than the greedy algorithm at any other value of r. When $r = 0$, paths having fewer steps exist, but we cannot find them; when r is large, the best route to take is unlikely to be much shorter than a path you'd take over strictly local links. Freeways are not good for traveling only a few blocks.

Power Laws

Sparseness, clustering and small diameter are not the only distinctive properties of large real-world graphs. The degree sequence — the number of vertices with each possible number of edges from 0 to $n - 1$ — is also important. You know the degree sequence of a lattice is simple. All vertices have the same number of edges, so plotting the degree sequence for all the vertices forms a single sharp spike. Any randomness in the graph broadens this peak. In the limiting case of an entirely random graph, the degree sequence forms a Poisson distribution, falling off exponentially away from the peak value on both sides. Now the probability of finding a vertex with k edges grows negligibly small for large k because of the exponential decline.

Real graphs such as the World Wide Web graph behave differently. A power law describes the distribution of degrees and not an exponential. That is, the number of vertices of degree k is given not by e^{-k} (an exponential) but by k^{-g} (a power law, where the power g is a positive constant). The power-law distribution slopes away more gradually than would an exponential. It is this drop off that permits highly connected vertices of very large degree. We discuss graphs of a closed and an open circulation in their respective chapters.

B. MODELING

Models and Modeling

Models represent nature symbolically. Models simplify important relationships making nature tractable. An effective model is often easier to test and analyze than is a physical system, as the model responds, within limits, in the same way. Models can be mechanical, morphological, physiological, mathematical (that includes dynamical systems models), finite element models and geometrical models.

Traditional Models

Traditional models are of wood, clay or remain as sketches and blueprints. For large objects, models are smaller and easier to evaluate before committing to construct the real thing. Today's models, however, are analytical and abstract in contrast to a physical model that is literal and concrete. First a beautiful early example of a model of a closed circulation.

Harvey Circulation Model

William Harvey discovered the circulation of the blood and published his treatise on the motion of the heart and blood in 1628. Harvey did not trace the connections between the arteries and the veins, nor did he follow blood around the circuit. Harvey never saw how arteries and veins connected, but his intuition told him they did. Malpighi finally observed capillaries in 1661.

Harvey was a systems modeler. He demonstrated that circulating blood had to be a necessary logical consequence of his observations. Harvey's reasoning is important as it demonstrates how model makers think. Harvey's first question was: Does the blood move the heart or does the heart move the blood? Because with each beat Harvey had seen the heart grow hard enough to resemble a contracting muscle as blood left it during the squeeze phase, called systole, and he had then seen that the heart then softened like a relaxed muscle after blood left it, Harvey reasoned that the heart had to be the prime mover in the system.

Harvey studied next the anatomy of the valves of the heart and showed that the valves would permit blood to flow in just one direction. In the living body then Harvey ligatured arteries and veins and saw blood accumulating on one side of the tie and draining away from the other side. Harvey measured the differences in volume between the dilated heart and the contracted heart and calculated that within an hour the human heart must pump some five hundred pounds of blood into the arteries, a volume that exceeded the weight of the whole body. Obviously, the only hypothesis that made sense of Harvey's observations was that the blood pumped into the arteries had to return to the heart and that this could only be happening through the veins. It, therefore, followed that arteries and veins had to connect somewhere, and that blood circulated through the body.

The major transport functions of mammalian blood are to move molecules of digested food from the gut to the rest of the body, to transport wastes from the rest of the body to the kidneys, to transport respiratory gasses between the lungs and the cells of the body, and to transport hormones from where they are made to where they have their actions. Blood, of course, also serves in hemostasis, combating infection, heat distribution and a host of other functions. As we shall see shortly, insect blood does not transport respiratory gases.

Modeling: Syntax and Semantics

We are most concerned here with mathematical modeling. To construct a model of something as Harvey did, we construct a representation that then substitutes for the real object. Model theory, a branch of mathematics, plays a role in logic that is similar to the role arithmetic plays in the rest of mathematics. However, beyond the basic ideas, advanced model theory is its own specialty.

In applied mathematics, a model abstractly represents some material reality taken from the world. This idea differs from how logicians use the idea of model. To them a model of a theory is something concrete, as much as any mathematical object can be concrete, and the model is the object that ultimately satisfies all the axioms of a theory. Logicians use formulas or sequences of symbols to write the axioms. Rules always determine formation of these sequences; these rules are syntax. Model theory presupposes we interpret the symbols in the formulas in some defined way, so that our interpretation is what brings meaning to our formulas. Our meaning ultimately translates into true or false statements about our observations. This interpretive notion is semantics.

What's important is that modeling occurs more at the level of semantics and uses little syntax. In building models we try not to learn what a specific structure we superimpose upon nature is, but we want to learn what is true about our model of nature and how we can prove this trueness or validity, referring back to the whole universe of our ideas, even though some of these ideas may as yet be inaccessible. In this way, our modeling contrasts with theoretical computer scientists' study of algorithms. Algorithmic studies are mainly syntactic.

A model of a bee's circulation becomes the basis for visualizing and generalizing a geometrical process. Because the model is what gives shape or form to our thinking, we begin by classifying and comparing our information or observations. Only later might we employ the implication symbol to indicate the formalization of the system that ultimately would grow into an axiomitzation formulating the rules for proof. If our model is effective, the model is easier to test and analyze than would be studying an actual bee, because within the limits we place on nature that we define by our observations, our model can respond in the same way we would expect a real bee to behave.

To model something as complex as the bee's circulation requires combining visual and analytical thinking and frequently the use of metaphor. To simulate a shape in a model is to define a shape that will come to house a process. Where we were once satisfied with two-dimensional graphs as being adequate representations, we now demand topologically valid and analytically complete models having three dimensions that change with time. Analysis then reduces our measurements and simplifies our 'data' into manageable concepts.

Modeling is then a loose integration of mathematical methods taken together to describe a shape or a process, often in terms of an appropriate metaphor. Computer aided geometric design (CAGD), for example, applies the mathematics of curves and surfaces usually employing the parametric equations of differential geometry. We use computational geometry to design and analyze geometrical algorithms.

Modeling Must Be At Several Levels

In modeling a circulation and then shrinking it, we must use a model, because a bee is too small, too complex, and time cannot be controlled. Using the rendering capabilities of computer graphics, we can explore more functional qualities and then change these at will, but still the problems and models continue to be too complicated, so ultimately a working rendition must come down to multi-scale modeling.

Approaches employing traditional, mono-scale modeling have proven themselves to be inadequate, even using large supercomputers, because the ranges of scales and the large number of variables involved are computationally prohibitive. Thus, there is a growing need to develop systematic modeling and simulation approaches for multi-scale problems. We have made some progress. For example, we can compare local changes in patterns of blood flow within a web of vessels with changes occurring within larger portions of the circulation as seen globally. Such a heterogeneous model uses Navier-Stokes equations to describe the three-dimensional flow within a single artery. We then couple a model of this flow to a systemic, zero-dimensional model of the whole circulation. Using this geometrical multi-scale strategy, we have joined an initial boundary value problem to an initial-value problem to predict a change wrought upon a circulation by surgical intervention, such as might occur following blood loss and onset of hypovolemic shock. Such involved models suggest we can obtain useful 'meta' information by matching conditions prevailing in two sub-models within a single numerical simulation.

Multi-scale Modeling

By its nature, multi-scale modeling is highly interdisciplinary. It has evolved independently across many fields, because a broad range of scientific and engineering problems involve multiple scales. Even though multi-scale problems are extremely difficult we are making progress. Inter-discipline communication difficulties, however, are rampant. Even though circulatory physiology and oceanography deal with currents and flows, many findings remain confined to the disciplines in which they were first studied. Much of this has to do with scale differences, but different groups think in different jargons. The overarching question again is how does one reliably compute larger reliable coarse scales while at the same time accurately modeling the net effect smaller events and structures have at subgrid-sized scales in both materials and fluids? (Ref: Multiscale Modeling).

Topology

We need to understand topology as well. How are simple elements joined, and how do we preserve these attachments when we transform the model? Topological properties are not metrical; they concern connectivity and dimensional continuity. These are the things that when *transformed* stretched, bent, twisted, or compressed without tearing, puncturing or inducing self-intersections still persist unchanged. Much of what the hemocoel does has to do with the topology of closed paths, piecewise flat surfaces, and closed curved surfaces. Ultimately physiologists and biologists will have to acquaint themselves with such formulations as the closed-path theorem, and the Jordan curve theorem.

Simulation System Level Modeling

MEMS devices contain varied components including electronics. We can model mini- and micro-scale MEMS using classical mechanical, electromagnetic and thermodynamic theory. Most studies concentrate on designing, modeling, and fabricating these systems. Comprehensive analysis precedes prototyping and fabrication. A common representation that encompasses multiple energy domains becomes useful in modeling the whole system. The bond-graph notation, based on energy transport (or power flow) may represent an entire system at the highest level. Ultimately, one seeks to know the dynamical behavior of the entire system.

But most transducers are nonlinear; they involve at least two energy domains, and they operate in the large signal regime. Direct numerical simulation of the dynamics of the fully meshed distributed model of such a system is computationally difficult and expensive. Therefore, one needs to reduce the degrees of freedom from hundreds or thousands in the meshed 3-D model to as few degrees as possible. We can then use such reduced order models to simulate and approximate the dynamics of the whole system. Such macro-models, however, should agree with our 3-D numerical simulations and our experimental results when describing the macro behavior of the system. Macro-models can also represent the behavior of a subsystem in one energy domain as well as the interactions from other domains. Hence, we need to automatically generate macro-models, and then we must insert these smaller models into some system-level dynamic simulator.

We also need to develop procedures to make quantum models of nano-scaled systems. Such models should avoid the complexities posed by the many-electron wave functions of classical quantum mechanical formulations. The complexity of the Schrödinger equation describing even a six-electron carbon atom requires visualizing a six-dimensional space. Each added electron requires adding an additional dimension.

CAD

Computer-aided analytical tools generate methods for designing MEMS eliminating some expensive and cumbersome building and testing. We need accurate simulations: accurate both in representing a structure's geometry and its underlying constitutive properties. We also require behavioral models for the device as well as its electronics. Most devices are transducers. Transducers operate in multiple energy domains and strongly couple among these domains (e.g., coupled electro-mechanical and fluid-structure interaction). Transducer-like devices require specialized analysis together with accurate 3-D simulation. Again, designers of these things have diverse backgrounds, and the field is multidisciplinary. Because not everyone is conversant with everyone else's tools and procedures, we need to develop more user-friendly analytical tools. For instance, should electrical engineers need to perform finite element analysis of a mechanical structure or should mechanical engineers need to simulate an electronic circuit, they should have suitable interfacing tools.

Coming Attractions

Circulatory systems contain and direct flows of liquids and gases through organisms and devices. Flows confine themselves to either closed systems of tubes and pumps or move freely within open cavities. After introducing pump-tube anatomy in Chapter 4 and the bee's body in Chapter 5, we study the open cavity circulations of bees and insects in Chapter 6.

Chapter four

YOU CAN'T SHRINK A WOMAN

The Impossible Shrinking Woman

Even with what Gulliver learned in Lilliput and what Hollywood says, we still can't shrink a person down to the size of a bee. Even theoretically! Why? People are vertebrates, and vertebrates have closed circulations. It is this closed circulation among other things that prevents shrinking a warm-blooded vertebrate much smaller than a hummingbird and a cold-blooded vertebrate much smaller than a guppy.

Closed Circulation Basics

A closed circulation is a heart-pump and its blood vessels: arteries, arterioles, capillaries and veins. Blood circulates inside a closed system and rarely leaves the vessels. Arteries convey blood away from the heart to capillary beds, and veins convey blood back to the heart from the capillaries. Arterioles are stopcocks interposed between arteries and capillaries. The nervous system opens and closes the stopcocks. Acting together, these valves aided by the beat of the heart maintain central blood pressure in the arteries of the circulation, but individually, each stopcock drops the perfusion pressure in the capillaries and controls how much blood runs off into the capillary beds of its region.

In seventy years, your heart beats around 2,759,400,000 times (75 beats per minute) squeezing 400 million liters of blood around your circulation. Your circulation matches nutrients and oxygen to what your cells need, while at the same time, your circulation removes wastes, regulates your body temperature, and distributes hormones.

Systole and Diastole

Your heart-pump cycles through two phases all your life: systole and diastole. Diastole is the resting phase between beats. During systole, your ventricles contract and squeeze blood into your large arteries. Some of the blood from each squeeze distends your elastic aorta and pulmonary artery. The rest passes through the systemic and pulmonary vessels. As systole draws to a close, and the blood pressures in your ventricles fall below the pressures in your aorta and pulmonary artery, the aortic and pulmonic valves between these arteries and your ventricles close, preventing regurgitation of arterial blood back into the ventricles.

During both systole and diastole, blood returns from your veins to your heart. During systole, this blood collects in the atria. It enters the ventricles when the valves between the atria and ventricles open after systole. When your heart muscle rests between squeezes, the arteries supplying your heart muscle, the coronary arteries, supply blood to the relaxed muscle of the heart. During diastole, because the aortic valves between the heart and arteries are closed, recoil of the stretched walls of large arteries drives blood through your body's vessels. If you didn't store energy in the stretching of your arteries each time your heart pumped, no blood would flow to your lungs and body when your ventricles were filling. Remember: All your diastoles together are a bit more than half your life.

Stroke Volume

When you are not exercising, each beat of your left ventricle ejects about seventy milliliters of blood (the ventricle's stroke volume) into your systemic circulation. Your systematic circulation comprises many regional circulations in parallel that supply your brain, heart, kidneys and other organs. After blood in your capillaries gives up its oxygen and nutrients and picks up carbon dioxide and wastes from your cells through the walls of the capillaries, the same stroke volume of seventy milliliters returns to the right atrium and the right side of the heart through your large veins. With your next beat, the blood we've been following passes through the right side of your heart to your lungs where it will give up the carbon dioxide from your cells and acquire oxygen. This blood will then pass to your left ventricle and be ready for the next systolic squeeze to your body.

Cardiac Output

How much blood a heart pumps in an interval, usually a minute, is the cardiac output. Cardiac output is simply the heart rate times the stroke volume. Stroke volume is the volume of blood the left ventricle ejects into the aorta each beat. The right ventricle ejects the same volume into the pulmonary circulation. Seventy-five beats a minute times seventy milliliters per beat equals a cardiac output of five thousand two hundred and fifty milliliters each minute. Cardiac output varies from high outputs of ten or more liters during exercise to five liters or less at rest.

Matched Outputs

Each avian or mammalian circulation, then, comprises two pumps in two circuits that connect through the heart. The right ventricle pumps blood through the lungs under low pressure. The left heart pumps blood through the body under high pressure. Each pump's output matches the output of the other almost beat for beat. If the right and left heart pumps were not evenly matched, one might end one's life having all one's blood in one's chest or one's body.

Flow Equals Pressure Over Resistance

Each heart pump creates a head of pressure that drives flow through its side of the circuit. Flow, pressure and resistance in each circuit are in ohmic relationship analogous to current, voltage and resistance in electrical circuits. Flow equals the mean difference in pressure between the arterial and venous ends of each circuit divided by the total resistance of the circuit. The quotient between the pressure head and the regional flow resistance determines how much blood flows in each circuit. The body provides much more resistance to flow in the systemic circuit than the lung resists flow in the pulmonary circuit, accounting for the differences in pressures and the muscularity of the ventricles, the right ventricle being thinner and less muscular than the left.

So in review, the left heart ejects blood into the systemic vessels supplying the body. These vessels form many regional circuits serving liver, brain, muscle, bone and other organs each separately controlled, specialized in design, and arrayed in parallel. The right heart pumps blood under lower pressure into the vascular beds of the lungs. These beds, more uniform in design, are where oxygen and carbon dioxide exchange between blood and air in the alveoli across the walls of pulmonary capillaries.

Size Differences

The difference in sizes of insects and vertebrates accounts for their very different circulations. Insects range from about a thousandth to a tenth of a meter long, while terrestrial vertebrates range from about mouse size up to five meters. There is very little size overlap between these two groups. Thus, very few insects are as large as the smallest birds or mammals.

A large difference in the number of species — estimated to be of the order of more than a million for insects and twenty thousand for terrestrial vertebrates associates with this size difference. Additionally, insects have short life cycles, and there are many more of them. Insects are very successful at remaining small in air on land and avoiding water loss, as the surface to volume ratio of insects is very high. Vertebrates, on the other hand, have a much lower surface to volume ratio, and they successfully support large bodies in air on their internal mineralized skeletons.

Size and Temperature

Cold-blooded animals can be smaller than warm-blooded ones. It takes less energy and fewer energy stores for one's body temperature to follow passively the temperature of the surroundings than it does to warm or cool a body in an ever-changing temperature gradient. The smallest cold-blooded vertebrate is the stout infant-fish (*Schindleria brevipinguis*). The adults are paedomorphic, meaning that the adults retain larval organs and functions. Males are sexually mature when just seven millimeters long. Females are mature at about eight and a half millimeters. The males of the dwarf goby from the Indo-Pacific are about eight and a half millimeters and the females are about nine. These fish are about at the limits of smallness for a fully developed cold-blooded closed circulation.

Being small and warm-blooded on land, on the other hand, demands capacity for heating and cooling and, hence, larger expenditures of energy. Today's smallest shrew weighs about two grams. Based on measurements of a fifty-million-year-old jawbone from *Batodonoides*, an early insectivore, this animal may have weighed only one point three grams. Somewhere in this range then is the smallest warm-blooded body that can be supported in air.

Limits

In general, heat loss or gain increases as one's surface to volume ratio increases. As storage space for on board energy supplies dwindles, a limit arises, because to maintain a constant body temperature against increasingly transitory and unfavorable external gradients, energy must be expended. What energy cannot be stored must be consumed and used immediately. Shrews forage day and night.

Hummingbirds

Small and warm, hummingbirds are about at the size limits for a terrestrial closed circulation. The smallest hummingbird, the Cuban bee hummingbird, is about six centimeters long and weighs less than three grams. Ruby throated hummingbirds weigh about three grams and are about nine centimeters long having resting body temperatures around one hundred and five degrees F when beating their wings forty to eighty times a second as they burn large quantities of energy in rapidly contracting fast red muscles. A Ruby's resting heart rate is around two hundred and fifty beats a minute, and it breathes at about the same rate. But a Ruby's heart rate rises to twelve hundred beats a minute when feeding on the wing. How does the little heart do it? (Ref: Hummingbird and Shrew Energetics).

Large Hearts, Small Chests, Cold Nights

Remember birds' hearts are like ours: a heart of four chambers. The right side of the heart receives deoxygenated blood from the body and pumps it to the lungs while the left side receives oxygenated blood and pumps it through the body. A hummingbird's large heart is about two and a half percent of a hummingbird's body weight, and its blood contains hemoglobin and a large number of red cells for transporting oxygen. Because the bird's surface to volume ratio is so very large and its energy needs so very high, to be able to sleep without feeding all night, a hummingbird's body temperature falls from a daytime norm of one hundred five degrees F to an overnight low of about seventy degrees F. This nocturnal torpor allows a hummingbird to slow its heart and respiratory rate during sleep and thereby lower its basal metabolic rate and thus sleep through the night without feeding.

Cardiac Limits

In shrews and hummingbirds, small hearts supply energy to fast moving muscles that work almost continuously. Because these animals consume so much energy, to be awake is to eat. Their hearts beat so fast because small pumps can eject only a small stroke volume. If the heart rate is too fast, however, diastolic filling time may shorten, so much that the ventricles cannot fill and the coronary vessels cannot supply the heart muscle. If a heart spends too much time contracting in systole and cannot compensate in diastole, it can fail. Insects avoid this uniquely vertebrate problem of high output heart failure with an open circulation.

Circulatory Systems

Let's extend our general idea of a circulation to include devices, vertebrates and insects together. All circulations solve supply and demand problems: how to distribute flow to points where it is needed, and when supply and demand are both high, to match an area's need to appropriately increased flows. Needs and flows constantly change. For example, in a running vertebrate, blood shunts to muscles at the expense of the intestines. After eating, blood flows to our intestines limiting the supply to muscles. This is why, after dinner, so few of us push back from the table to run the mile.

Blood Supply to Organs

Before turning to open circulations, we consider how blood moves through the capillaries inside the three-dimensional volume of an organ. We often express blood flows to organs as volumes of blood flowing per minute into or out of an organ as a percentage of the total cardiac output. We can also calculate blood flow per unit mass of tissue. Both these values indicate where flow goes as a percentage of the cardiac output, but these simple numbers do not reveal the pattern of how blood traverses capillaries. As a rule, simplified global descriptions are inadequate to explain happenings at microscopic levels of resolution. In the heart, blood flow per unit mass of tissue is not uniform in space. Nor is blood flow constant in time over periods longer than a few beats. This finding is important in medicine where radiologists may perceive defects in deposition of thallium in x-ray films of coronary flows and attribute these to pathology. We may do surgery to repair a normal heart. What is abnormal and what is normal? Often we cannot tell.

Normal Blood Flow Through Trees

Blood flows are not uniform at the microscopic level. How vessels are arrayed can create non-uniform flow patterns. Tissues differ as to placements of their capillaries. In cardiac muscle, capillaries lie in parallel arrays alongside the muscle cells, and a single straight capillary may extend for centimeters. In heart muscle, many arterioles feed into one large dense network of two to four thousand capillaries per millimeter of cross-sectional area of tissue. Cardiac capillaries appear ideally situated to deliver oxygen and nutrients rapidly to active cardiac muscle. Other capillary beds in other organs have different arrays.

Trees

Vessels of a closed circulation are arrayed as trees. Large vessels and airways branch like tree trunks into smaller and smaller divisions leading to twigs. Each vessel 'tree' is a series of segments of cylindrical tubes that join at their dichotomous branch points so that, like twigs, down stream daughter branches have smaller diameters and are shorter than their parent vessels.

Swamps of Capillaries

Capillaries are the exchange vessels and have the smallest inside diameters of all the vessels. However, there are so many capillaries that their combined inside diameters are greater than the inside diameters of the arteries, so all arterial blood can pass through capillaries without pooling. Arterioles reduce the perfusion pressure of arterial blood to the lower blood pressures of capillaries.

Blood flow in capillaries is very slow and is not uniform. In one capillary, flow may slow down, halt and then reverse like water surges in a swamp. Oxygen, carbon dioxide and metabolites exchange through the walls of capillaries between the slowly moving blood and the stationary tissues. The differences between the ratios of the resistances to flow in the arterioles and the post-capillary vessels determine intra-capillary blood pressures and thus how much blood flows through the capillary beds to perfuse the tissues. Because capillary walls are thin, individual capillaries may swell or collapse depending on the differences in pressures across the capillary walls and flows within and outside them. High external pressures can collapse capillaries preventing adequate perfusion of tissues. A single capillary or bed may open or close depending on the needs of nearby cells. Capillary walls often dilate or constrict responding to locally produced, vaso-active chemicals. The small diameters of capillaries permit their thin walls to exert leverage upon high internal pressures (Laplace's Law). Large capillary surface to volume ratios facilitate rapid exchange.

Exchange

Some capillary walls have holes in them, but others are tight. Water-soluble molecules such as glucose traverse pores in the capillary walls, but lipids pass through the epithelia of capillaries direct. Non-fenestrated capillaries permit large polar molecules to cross with difficulty. Gaps between epithelial cells in sinusoids or discontinuous capillaries in spleen, liver and bone marrow permit even cells to pass.

What We Don't Know

Obviously how blood and materials move from moment to moment at the cellular level of resolution is unknown. One often presumes that high flow through a sequence of vessels suggests that the cells these vessels feed must have high rates of metabolism, but there is little data, even though whole organ data support the idea that increasing metabolic demand in tissues in turn increases local blood flow. Suggestions abound that local flow, transport capacity and metabolism match up in some way, but how these variables match up, if at all, is unclear. If regional feedback between cells and nearby capillaries controls local flow, how are these local regions coordinated together and regulated higher up? Such questions so far remain unanswered.

A Fractal Match

Organisms are fractal assemblies but only over a limited range of scales. In the simplest picture, we might imagine any animal to be a solid body of cells adjoining a liquid phase across a common fractal border. Animals with closed circulations especially exhibit fractal patterns of their tissues and vessels.

Fractal structures intricately intertwine stationary catalytic or cellular surfaces against a moving liquid and can maximize the area of contact. This arrangement maximizes apposed surface and facilitates optimum exchange across the barrier and turnover. This general almost fractal pattern occurs at many scales and sizes. Intricate interlacing of a stationary catalytic system, the cells, with a pumped liquid phase, the blood or hemolymph, minimizes distances and resistances to transport down to even sub-cellular scales. Such fractal anatomy of interfaces optimizes turnover through all levels of organization.

Using a similar concept we may model both closed and open circulations together. After we have discussed open circulations, we shall compare transfer along this type of border with transfer through a percolation system. We may already apply concepts of percolation to the gel states of polymers on the cellular or molecular scale (Ref: Fractal Gels).

Matching Flows Across Exchange Surfaces

Examples occur in any organ, but the human lung is a "best" example, as air flowing to and from alveoli must match blood flowing through the capillaries around the alveoli. At large and smaller resolutions we can visualize the patterns of these distributions graphically, and using our graphs we can create our models. But first what constitutes flow matching in the lung?

In a closed circulation supplying an organ such as a liver or lung from a blood supply entering and leaving the organ through a single portal, one problem of distribution is how to divide up the flows internally so that all regions of the organ receive adequate blood. Liver or heart muscles seen at one resolution contain cells that occupy the volume inside the organ almost homogeneously. Blood flow backs up rarely. The alveolar surface of the lung shows a similar match-up, but because we can easily visualize our lung as a two-dimensional surface, we shall use the lung as our example of a flow matched to need situation.

The Lung

The lung is an exchange surface of many flat cells. The surface, about the size of a tennis court or about one hundred and thirty square meters, has the form of three hundred million grape-like sacs or alveoli spread over one side of a thin membrane on the other side of which courses a bed of capillaries. Gases must traverse this alveolar-capillary membrane as they cross in both directions between alveolus and capillary. Our lung surface is crumpled within our chests, each folded half court taking up about five liters of space. Networks of capillaries surround each alveolus.

The capillaries around all our alveoli together must receive blood as evenly as possible, so that the alveoli high up in our chests and others lower down near our diaphragms all receive the blood they need to match the air they get: no more and no less.

So one-half of our distribution problem is how to divide up the stream of blood coming into the lung efficiently so that blood reaches all points of the combined alveolar surface evenly in approximately the same time. The other half of the problem entails getting air uniformly to the alveoli. Alveoli, no matter where they are in the chest, high up or low down, must receive adequate fresh oxygen from each inspired breath. Airflow and blood flow must match so that blood in some capillaries does not meet poorly ventilated alveoli and return to the heart empty-handed and also so that well ventilated alveoli receive more blood than poorly ventilated ones. So the goal of a matched supply system must be that ventilation and perfusion of the exchange system must match up almost perfectly over the area on both sides of the alveolar capillary membrane.

How We Match Flows

Vertebrates solve the ventilation-perfusion flow-matching problem by having the blood vessels of the lung and the airways as trees with their roots in the heart or in the upper airways extending down from the nose and mouth. Twigs of both of these trees intertwine and touch at the level of the alveolar-capillary units. Look at this match from the airway side. Air enters the trachea that soon divides into bronchi. The bronchial tubes branch and narrow their diameters over and over about twenty-three times until reaching their last division: the alveolar level. Each alveolus, with it skein of capillaries around it, behaves as a unit. However, shrinking the branching trees of airways, as with shrinking a simple pump-tube distributing system as we did in Chapter 1, would lead to blockage when the internal diameters of the airways grow too small so that flows diminish.

Look at the match-up now from the circulation's point of view. Blood leaving the right ventricle of the heart passes through pulmonary arteries and then analogously branching subdivisions of arteries all the way down to the pulmonary capillaries. The functional result of this design is that all points on the exchange surface of the lung are almost the same distance from the source of fresh air and blood: the main airways and the heart.

Tree distributions permit air flow and blood flow to remain almost entirely laminar and non-turbulent as they flow in very close proximity to each other on either sides of the thin alveolar capillary membrane. Lack of turbulence permits the work of pumping and the work of ventilating the alveoli to remain minimal.

Representing a Closed Circulation

A scale-free graph best represents the vertebrate circulation. In this graph, we see the circulatory network as connecting vertices or nodes, or sites of metabolism. These active vertices connect with each other through edges that are the actual blood vessels. The edges are directed as blood flows in one direction (away from the heart in arteries and returns to the heart through veins), so we distinguish incoming and outgoing links for each node.

The probability that a node receives blood from k distal flows equals the probability that this same node feeds k different downstream flows. Both inflows and outflows have similar distributions, because blood is neither consumed nor created in most nodes. (An exception would be blood entering the veins from venous reservoirs.) $P(k) \sim k^{-\gamma}$ in equals $P(k) \sim k^{-\gamma}$ out. Gamma is the probability that two nodes are connected. This scale-free network also displays small world character, as any two nodes may connect along existing links by relatively short paths, for example, all blood leaving the left heart traverses the aortic arch.

The diameter of the circulatory network spanned by our graph characterizes the degree of interconnectivity within the circulatory network. We define diameter as the shortest path for blood flow averaged over all pairs of nodes. This shortest path indicates the rate at which information (blood) spreads throughout the graph. If all nodes were fixed, the diameter of a scale-free network would increase logarithmically as new nodes joined the network. The diameter of the network of a larger animal comprising a larger number of metabolic sites might exceed that for a smaller animal. A graph of log $P(k)$ versus log k for the vertebrate circulation has no well-defined peak. For large k the graph decays as a power law, $P(k) \sim k^{-\gamma}$, appearing as a straight line with slope $-\gamma$ on a log-log plot, because the heart, lungs, kidneys and brain each have a large number of links. As happens in the World Wide Web, important nodal places (hubs) are vulnerable to attacks on the system. A heart attack or brain infarct can knock out the entire mammalian circulation. This single knock-out opportunity does not exist for an open arthropod circulation.

Chapter five

BEE'S BODY

Introduction

An adult worker honeybee represents the generalized insect body plan and its subsystems. However, this worker bee blueprint modifies the plan of a generalized insect. This chapter summarizes bee anatomy. After laying out her plan, we shall shrink her dimensions.

Why choose honeybees? Honeybees are large common denizens of gardens, are easily raised, are not yet extinct, and we understand honeybee anatomy and physiology quite well compared with these in other insects. Evolution over geological time has perfected myriad variations on the 'insect' theme.

Insect bodies are compact, light and stable. They function well in all climates and at atmospheric pressures. There are at least five to ten million species of insects and only forty-two thousand vertebrate species; consequently, insects form an incredibly rich but largely unknown source of ideas to be mined for miniaturizing our engineering innovations. We shall continually ask ourselves: how might a shrinking bee generate ideas for human designers of small mechanical devices.

Crayfish and lobsters and other arthropods, also constructed on 'bee-type' plans, are much larger than insects, so their anatomy appears more comprehensible to our unaided eye. Arthropods come in large as well as very small sizes indicating the arthropod body-plan remains relatively unaltered and, hence, quite recognizable and robust over a very broad range of sizes. Evolution of external arthropod armor permits a tank-like defending shell as well as a secure attachment for muscles and water-proofing. Later modifications that came to be wings also now serve as solar heating panels and signaling devices.

Warning

Keep in mind however, we should never consider insects, spiders and other invertebrates to be 'just' shrunken vertebrates. We often base our human assumptions about how 'animals work' on what we know of human biology. For example, a 'law' based on mammalian pulmonary and cardiovascular systems derives a three-quarter power allometric scaling relationship for metabolic rates of organisms describing distribution of gases and foodstuffs through linear space-filling fractal networks of branching tubes suggesting that tubular distribution systems characterize all organisms (Ref: Allometric Scaling). Because invertebrates often are so very small, they cannot employ closed systems of pumps and tubes in a vertebrate configuration.

From Precambrian Chains to Three Segments

The earliest arthropod bodies undoubtedly formed from worm-like chains of similar tubular segments in Precambrian sediments. So we must first inquire how might a wormy chain of segments have become the plan for the insect bodies we have today.

Imagine a centipede-like body progressively changing perhaps stepwise into a bee's body. A related question might be that if change arises from mutant genes, how were these early bodies able to tolerate large-scale mutations in the genes determining their form without the bodies developing suddenly some grotesque abnormality and becoming dysfunctional enough to end the evolutionary sequence?

Hox Genes Keep Legs Off of Heads

Mutations that change body parts, if the changes are too extreme, might kill their bearers before the bearers of these new mutants achieved reproductive maturity. Enter *Hox* genes. *Hox* genes are 'super' genes that apparently reside at the top of a hierarchy of genes to control many other genes lower down.

All of an animal's genes, working together, ultimately determine the animal's form. One gene, having the beautiful name of ultrabithorax for example, regulates where limbs form on the body. Usually limbs do not grow on heads or on abdominal segments. In insects this gene, shortened to *ubx*, suppresses formation of limbs and wings in the abdominal segments and also ensures that wings and legs develop only in the thorax. We know this because if we remove the *ubx* gene from fruit flies, legs sprout from every abdominal segment. *Ubx* thus suppresses abdominal legs. Then if we place the *ubx* gene inappropriately into the thorax where it is not usually active, *ubx* suppresses development of legs where legs normally occur. Thus, *ubx* genes turn off and turn on genes that more specifically and minutely control development of the complex structures of the legs and wings. If we take a *ubx* protein from brine shrimp — these crustaceans carry legs on their abdominal segments — and we place this brine shrimp protein into the thorax of fruit flies, only fifteen percent of the limb development of a fruit fly's legs gets turned off or suppressed. This change suggests that over the interval of geological time between when brine shrimp arose to when fruit flies emerged, the *ubx* gene evolved and even changed its function.

These and other genetic findings suggest that altering just one small part of a tightly organized genetic system may, in turn, modify a much larger pattern. Not only are individual genes not immutable, whole patterns of genes are not fixed through time, as they can adjust their collective responses to evolutionary demand. It would be useful if human designers could develop devices having at least some of the sophistication of insect bodies, so that devices might evolve according to the demands of their changing environments.

Protein Plus Silicon

Protein joined with silicon is a step in the right direction. We have taken a major conceptual step. We can now link enzymes that are biological proteins to a chip-based technology inside a three-dimensional anatomical array or skeleton. Biological molecules that couple to electronic circuits open up possibilities for DNA wires and reactor cells containing enzymes.

One new technology now harnesses the energy from electric currents produced during photosynthesis. This energy after development may extend the lives of batteries in cell phones and portable devices (Ref: Energy from Photosynthesis). Early attempts to combine silicon and proteins failed largely because the proteins that capture energy from light during photosynthesis denature and change their form if they dry out. A peptide detergent, however, helped stabilize the protein, allowing it to channel energy within a chip. The researchers placed protein molecules, extracted from spinach, heads up while leaving spaces between each protein molecule on the chip. They then filled the spaces between the proteins with the peptide detergent, so through self-assembly, the protein component and the detergent combined. This array is sandwiched between layers of plastic, gold and indium-tin oxide, a transparent semi-conducting material. The chips are durable, and they continue to pass current for weeks. Researchers can repair individual chips by injecting more protein into the spaces if needed. In a similar way, we may pile up many similar layers and may then wrap these layers around conventional batteries to create a three-dimensional array that generates increased battery power.

What is interesting in this example is the anatomy. Precise positioning of a protein and silicon relative to each other as well as within a larger battery 'system,' creates a compact, organized unit that we may now also repair. The insect body has been layering and integrating itself between and within twisted and organized systems of layers since the Devonian. We have much to learn, but we may be on our way.

Hexapod Uniformity Allows Shrinkage

The most distinctive change insects made in the primitive chain of segments having legs on each, was by reducing their walking legs from many to only six. These six legs attach to a specialized, consolidated powerful thorax of just three body segments. As with smaller insects that resemble larger ones, the compressed anatomy of the thorax, originally most probably a minor variant, became standard in large as well as small arthropods. Additionally, the anatomy of the thorax is stable, as it appears to 'shrink' down quite readily to fit within the smallest bodies.

The components of the insect plan retain for the most part their customary relationships, suggesting that anatomical relationships are crucial to insect success and are robust enough to have remained little changed over evolution, as organs and their parts most often retain their standard positions. These observations taken together with the worldwide distribution of arthropods demonstrate how successful design of the arthropod body is. Would that we might emulate it and that the designs for our devices were as robust.

Body Plan in Three Parts

A bee's body parts like ours are often hidden. A fluffy coat of hairs obscures the outline of a bee's body especially in colder climates where hairy bees are the rule. We shall shave the hair to imagine a simplified common plan of an adult honeybee, all the while remembering honeybees lead very specialized lives, and

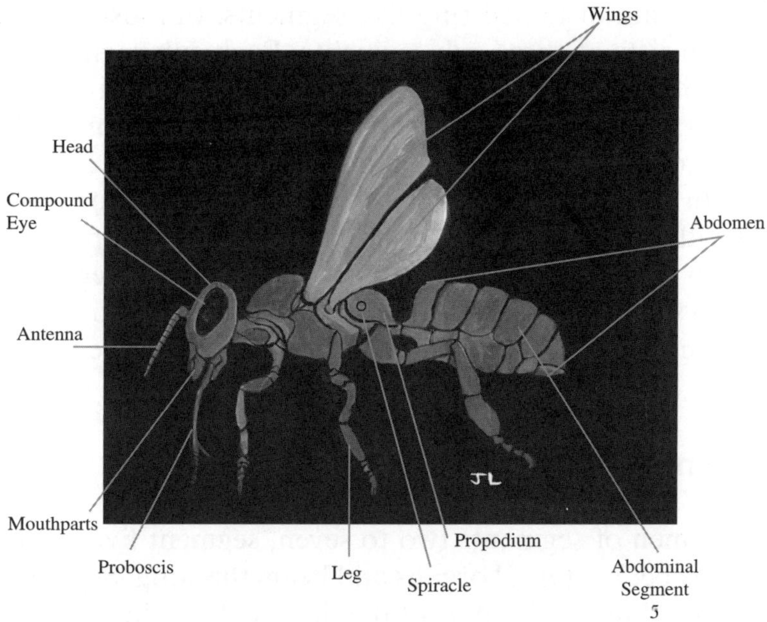

Figure 5.1. The Three Parts of the Bee's Body. Side view of a worker bee devoid of hairs. The thorax, unlabeled, holds the legs and wings.

therefore, have very specialized bodies. A worker bee's body is in three parts: head, thorax and abdomen.

Head

The head holds her eyes, her antennae and her mouthparts she uses for feeding. A slender flexible neck joins her head to her thorax.

Thorax

Her thorax houses the machinery of walking and flying. Her thorax and the third section of her abdomen or trunk arise from the primitive chain of joined ring-like segments. In most insects the thorax consists of three sub-segments. Each sub-segment holds one pair of legs, but in the bee, the thorax has four divisions, the prothorax, the mesothorax, the metathorax and the propodeum. A bee's propodium is the first abdominal segment of most other insects. In bees this propodial segment has moved forward and joined with the thorax. The prothorax holds the first pair of legs. The meso and the metathorax have each a pair of legs, but the latter two segments also hold and control the two pairs of wings. A short peduncle or stalk connects the thorax with the abdomen.

Abdomen

The abdomen of segments two to seven, segment five is labeled, houses the bee's internal organs and bears the sting. Segment one is the propodium and is part of the thorax. The sting of a worker bee is the modified ovopositer or egg layer of other insects.

Adaptations

A bee's body is modified for specific activities. Anatomical and physiological modifications are superimposed upon and integrated into her generalized body plan. For example, the feeding organs of bees contain similar parts as those of other insects, such as crickets, but a bee's mouthparts differ in shape and function in that they form a proboscis for ingesting nectar and pollen. The gut or alimentary canal of a honeybee is specialized for holding honey. Her respiratory system is enlarged accommodating rapid flight. Bee wings enable swift flight, but they can also sustain heavy, rapidly changing loads of nectar, honey and pollen. A bee uses her legs not only for walking. Her legs are shaped for holding on, feeding brood and young bees, for cleaning herself, and are adapted for performing other jobs around the hive, such as carrying pollen. The sting of a worker bee discharges formic acid and not eggs. Abdominal glands produce wax for the building of honeycombs.

Bee in More Detail

The cranium-like head arises from fusing several primitive segments together. The head carries four pairs of appendages, the antennae, the mandibles or the bee's jaws, the maxillae and the labium. In bees, these latter two fuse into the proboscis adapted for feeding on liquid nectar and honey. The head of a bee carries a pair of large compound eyes. Between these big eyes are three smaller eyes or ocelli. The head attaches to the thorax. The cavity inside the head joins the cavity inside the thorax, and the neck tube allows the esophagus, nerves, blood vessel, tracheae and salivary ducts to pass from the head into the thorax much as cables traverse a conduit. Internally, within the cavernous space of the head, two large bars of cuticle extend from the sides of the neck into the cavity supporting and strengthening the head.

Because the thorax contains, supports and supplies the powerful motors driving her wings and legs, muscles almost fill the cavity of the thorax. These muscles are the muscles of locomotion as well as those moving the head and the abdomen. Hemolymph flows over and around these muscles, as tracheae admit air from outside and convey oxygen directly to the cells of the active muscles. Functions of the thorax depend on mass, flight activity, and load, and especially complicated relationships between metabolic activity and size, and are highly regulated (Chapter 10).

Thorax muscles: Muscular systems in the thorax are tightly and conservatively organized, as any additional weight costs dearly in terms of fuel. Cut open a bee's thorax to see that it is almost completely filled with masses of muscle fibers. You can see these with the naked eye, but if you tease the muscles apart under a dissecting microscope, you will see that the muscles are compartmentalized into units. Each set of muscles has multiple functions. Half of the muscles in the thorax, a right and left mass, course down the center of the thorax from the head back towards the abdomen. These are the dorsal longitudinal muscles that attach to a complex set of hinges and upon contraction, depress the wings. Other muscles in the thorax forming almost half the total mass of muscle, course at almost right angles to this first set. These crossing muscles, the dorsal ventral muscles, are the elevators of the wings. Depressor muscles of the back also serve a double function as they also elevate the wings. Contracting depressor muscles of the wings elevate the back and turn the wings down. During flight and preflight warm-ups, the two sets of muscles contract and relax alternately (Ref: Flight).

Flight

The two wings on each side hook together during flight. Flapping up and down alone does not permit flight; but driving forces arise from propeller-like twists given to each wing during up and down strokes. Large nerve centers called ganglia in the ventral nerve cord, control the wings. Wings hinge by their narrowed bases to the thorax and are free to move up and down, but flight requires forward and backward motions as well as twisting or partial rotation of wings on their long axes. Thoracic muscles provide power, but most of these muscles attach not to the wings themselves, but to movable parts of the thorax that move the wings indirectly.

Abdomen

The abdomen contains the viscera: the stomach, intestine, reproductive organs, and the external genitalia that usually are concerned with mating and egg-laying. In worker bees, the genitalia are incorporated into the mechanism of the sting.

The ten segments of the larval bee's abdomen reduce to nine segments in the adult, as the first abdominal segment joins the thorax to become the hindmost thoracic segment, the propodium.

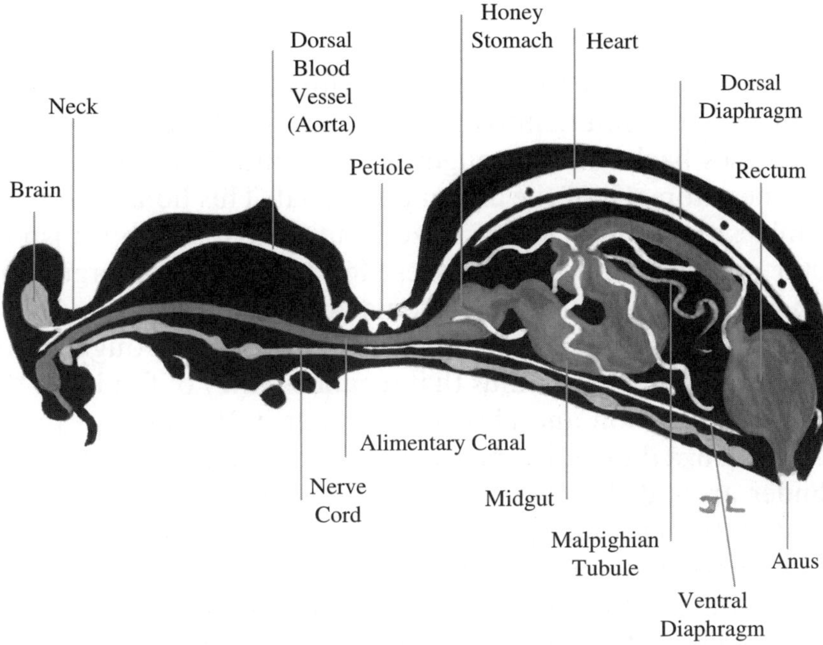

Figure 5.2. Bee's Internal Anatomy. Lengthwise section through a worker bee.

Each abdominal segment contains a large back plate, the tergum, and a smaller ventral plate, the sternum. The successive plates of sequential segments overlap from front to back. The plates connect with each other through inter-segmental membranes. The sides of the segments connect with each other through infolded lateral membranes. Our primitive sequence of joined-rings makes the abdomen distensible and contractile in length as well as allows the tip of the abdomen to bend up and down. Long retractor muscles that course along the length of the abdomen pull the segments together to shorten the abdomen. Shorter protractor muscles in each segment oppose the retractor muscles to lengthen the abdomen. Compressor muscles within each segment draw the tergum and sternum of each segment closer together. A short tubular stalk, the petiole (think of a wasp's waist), unites the abdomen to the thorax permitting much movement between the two. The nerve cord, alimentary canal and dorsal vessel traverse the petiole, again as in a conduit.

Alimentary Canal

A sucking pump in the head draws food, honey or nectar or pollen, from the mouth into the esophagus. The tube of the esophagus passes backwards through the neck and thorax to become the honey stomach of the alimentary canal. This honey stomach resembles the crop of other insects but is adapted to transport nectar or honey and to store these foods for later regurgitation or digestion. From the honey stomach food to be digested then enters the true stomach or midgut or ventriculus through a narrow muscular proventriculus that regulates entry of food into the stomach. Digestion and absorption occur in the midgut. Inside the stomach, a thin peritrophic membrane secretes a delicate filmy cylinder around the mass of food. From the wall of the stomach, digestive enzymes pass through this peritrophic membrane to digest the food. Later, the products of digestion pass back through the peritrophic membrane once more before traversing the stomach wall to enter the hemolymph. What remains of the meal then enters the intestine. The intestine is divided into a narrow anterior part that is coiled and a larger pear-shaped posterior intestine or rectum that opens to the outside through the anus. These latter structures serve to absorb water and to discharge wastes. Bees retain feces in the rectum until they are evacuated outside the hive. In an over-wintering bee confined inside a hive, the rectum may distend to fill a large proportion of the abdominal cavity.

Malpighian Tubules

These little tubes are excretory organs serving as kidneys. Malpighian tubules remove nitrogenous wastes and salts produced during metabolism from the hemolymph. A hundred or more of these thread-like tubes join the alimentary canal where the intestine joins the stomach. The free ends of the tubes extend for long distances within the abdominal cavity, and like hoses, can move about within the spaces surrounding the organs. The hemolymph that washes over the organs enters the free open ends. Wastes move along the tubes to discharge into the intestine and are eliminated with the feces.

Fat Bodies

Fat bodies are irregular masses of a soft white tissue composed of large, loosely joined cells that are scattered throughout the body cavity. Fat bodies are most numerous in the abdomen. The cells of the fat body contain oily fat. The fat bodies store products of digestion that the bee does not need immediately.

Respiratory System

For flight and other metabolic processes demanding oxygen, oxygen must in some way pass through the impervious cuticle to reach the mitochondria inside all the cells of the body. How do bees accomplish this?

Bees exchange gases through the tracheal system, a branching network of tubes throughout the body. Air conduits fill up large portions of the interior body. The tracheae open to the atmosphere through the spiracles, little openings resembling portholes in the cuticle. Each porthole possesses a closing mechanism that reduces loss of water from the respiratory tree. Oxygen influx and carbon dioxide efflux must occur through the tracheae, as the cuticle is quite impervious to these gases.

Spiracles

The adult honeybee has three thoracic and seven abdominal spiracles on each side, because the first abdominal segment, the pronotum, has joined the thorax. All the spiracles except the smaller second are capable of controlled closing and opening. Spiracles posses a muscular apparatus controlled by the nervous system that closes to prevent escape of air in the tracheal tree or to adjust the flows of air through the spiracles. Most insects have ten spiracles on each side, two on the thorax and eight on each side of the abdomen. The first and largest spiracles lie between the pro and meso thorax, each partially hidden by the overlapping edges of the pronotum. The second spiracle is small lying between the upper angles of the plates of the meso and meta thorax. The third spiracle, fully exposed, is on the side of the pronotum. The next six are in the lower portions of the first six tergal plates of the abdomen, and the last spiracle is at the base of the sting.

Tracheae

Internally tracheae form branching invaginations of the cuticle deep to the spiracles, the internal diameters of the tracheae growing smaller and smaller at each bifurcation. This system of bifurcations transports oxygen directly to where it is utilized without its passing through the hemolymph. The terminal branches of the tracheae are the tracheoles. The tracheoles end on or close to the cells. For example, up to ten percent of the mass of flight muscle may be air tubes. There is, always a tradeoff between filling a muscle's volume with muscle fibers or airways, as space on board is at a premium. Some of the smallest terminals of the tracheoles may even indent cell membranes, reducing the distances for diffusion to the mitochondria. Diffusion through these smallest air tubes is continuous with diffusion of oxygen through the tissues. At rest, tracheoles contain some liquid. During flight, however, this liquid is absorbed, so that now a continuous pathway of air supplies the increased metabolic demand of the mitochondria, as the diffusion of oxygen occurs faster in air than in water. The lengths of the paths for tissue diffusion, however, set an upper limit on how big tissues, organs and ultimately the size of an insect may be.

Air Sacs

Many tracheae along their courses dilate into thin-walled air sacs. These sacs expand to fill much of the volume at the sides of the abdomen and within the smaller spaces of the thorax and head. Some sacs, drawn out like filaments, even extend down into the proximal ends of the legs.

Ventilation

To ventilate the respiratory system, opposing sets of abdominal muscles contract alternately like a bellows, producing dorsal-ventral and lengthwise contractions and expansion of the abdomen. To move air through the spiracles, the air sacs must also expand and contract from the alternate compressions and expansions of the abdomen. Because the air sacs are collapsible, hemolymph can shift in the hemocoel as the intestinal tract expands and contracts with feedings. Air sacs may allow growth of organs without changing the exterior form of an insect's body. The tracheae themselves are more or less rigid, as they possess spiral thickenings of cuticle within their walls that keep them open. Hence, the tubes themselves do not respond much to increases and decreases of pressure around them, but the thinner air sacs do.

Gas Exchange

From the air sacs, the tracheae branch and ramify to the appendages and the organs. Tracheae then end in tracheoles that terminate blindly within cells. Proximally tracheoles are about a micrometer in diameter, but they taper to diameters of a tenth of a micrometer. Physiological cascades draw oxygen from this fluid into the mitochondria during oxidative metabolism. It is uncertain if the tips of tracheoles ever become truly intracellular.

Distributions of tracheoles to end-organs reflects their demands for oxygen, as tracheoles are most numerous in muscles, glands and neural tissues where oxygen consumption is highest. Some carbon dioxide produced by the metabolizing cells cannot be funneled into the air conduits. This carbon dioxide, instead, discharges into the hemolymph, and after circulating may diffuse through soft areas of cuticle or be converted into bicarbonate and excreted through the Malpighian tubules into the midgut (Ref: Respiration).

Might we someday construct air sac-like reservoirs in devices that could accommodate shifting fluids within rigid interiors? Sacs might also insulate hot spots in devices much as they help insulate thoracic motors by confining heat flows, permitting a warm thorax and a cooler abdomen to co-exist at different temperatures. More about heat exchange in Chapter 9.

Chapter six

CAVITY TRANSPORT

Overview of Chapter

Open circulations supply many invertebrates: from crabs, lobsters, insects and spiders to starfish, sea urchins and some mollusks. More animal species employ open circulations than the closed 'pump-tube' circulations of vertebrates. In open circulations, movements of the body assist open pumps to propel blood or hemolymph over and around organs inside an open cavity or hemocoel. From the pump or heart, blood traverses tubes passing direct to gills or brain, but ultimately blood leaves the tubes to percolate through the open cavity. Because open circulations lack capillaries, tissues and organs surrounding the cavity must take up nutrients and discharge wastes into the slowly moving blood direct. In some forms, the 'blood' of the open circulation transports oxygen in addition to foodstuffs and wastes, but in insects, oxygen distributes via a separate tubular system of tracheae. Open cavity circulations, unlike closed circulations, continue to function when shrunk to spider or gnat size making cavity circulations ideal models for supplying miniaturized microfluidic devices. We begin with a thought experiment.

THOUGHT EXPERIMENT

Efficiency

Efficiency is the ratio of the work a system performs to the energy expended in performing the work. To maintain or increase efficiency as a system shrinks, its decreasing mass and changing surfaces must keep pace with changing energetic processes. Integration and efficiency depend upon on a machine's anatomical design, and in living systems, physiology continually feeds back to fine-tune morphology. If a machine or organism is to adapt to changing environments, structure and function must evolve together. Structure is inseparably linked to function.

Surface-to-Volume Ratios

To appreciate how the need for a circulation relates to a system's size, be it animal or machine, we consider the system's surface-to-volume ratios. Compact efficient systems minimize extra space and weight. Efficiency of transfer of energy and materials increases when thin surfaces having maximal areas fold up compactly within minimal volumes. One example is how a hemocoel and tracheal system are coordinated, matching their surfaces. Circulating hemolymph moves a wall's thickness away but alongside air flowing through the conduits and sacs of the respiratory system that interlace the bee's body.

Model

Our model for a thought experiment generalizes needs of supply and demand to satisfy metabolism as size, activity and demand increase. When bodies are small, diffusion suffices.

Conceive of a spherical bee smaller than a gnat. For energy she requires oxygen and nutrients, and at the same time she must eliminate carbon dioxide and water as waste products. Because she is spherical, we assume all of her interior points of metabolism where she consumes oxygen and fuel and produces carbon dioxide and water to be equidistant from her body's surface. If she is small enough, her body equi-permeable to the molecules so that distances for diffusion remain optimally short, and if supplies of energy in the environment around her suffice, diffusion of both oxygen and food molecules from the environment through her body wall to her tissues as well as diffusion of wastes from her body to her environment will supply her metabolic needs. Diffusion suffices as long as diffusion time and path length remain short enough.

Diffusion Limits

It might take up to a minute for diffusing oxygen to penetrate to the center of a single amoeba floating in a bath, if we calculate that the average time for diffusion equals the diffusion distance (centimeters) squared divided by a diffusion constant that is about $\sim 10^{-5}$ for most small molecules in water. If our amoeba's cell membrane is one hundred Angstroms thick, then diffusion time might be 0.0000001 seconds, but for diffusion to supply a spherical animal ten centimeters in diameter, time to diffuse might take one hundred and twenty days (Ref: Diffusion Rates). But now what happens when body size and or activity increase?

Size Increase

Unassisted diffusion cannot supply adequate oxygen fast enough for diffusion distances greater than about two hundred and fifty microns, so pumping or convection must be added to speed transport. Let's increase by ten times each of our bee's three dimensions. Her weight, now a cubed function, goes up a thousand-fold. So for our bee now to remain as efficient as she was before we enlarged her, each minute she will need one thousand times as much food energy and oxygen, and during the same minute, she must excrete one thousand times the wastes including carbon dioxide she excreted before her size increased.

But how can she do this? Unaided diffusion can't suffice, because her once optimal distances for diffusion to supply all her central points of metabolism are now way too long. Also, if we wish her to fly and walk, she will have to increase her need for energy even more and will require a more rapid supply of fuel and removal of wastes. Our bee's marginally sufficient metabolism at rest now is insufficient for exercise, as her metabolic rate is tied to how rapidly she acquires oxygen and excretes carbon dioxide. Her rate of metabolism now restricts her activities and her mobility.

Surface Area Must Increase Relative to Volume

If our bee keeps her spherical shape, the surface of her body will have increased one hundred times, so ten times the original oxygen must enter her tissues through each square millimeter of her new surface area per minute. Similarly, ten times as much food energy must enter through each square millimeter of her gut each minute, and ten times as much waste must leave her body each minute. Therefore, to retain her present size, it behooves our bee to increase her surface area in contact with her environment relative to her volume; that is, she must increase her surface to volume ratio. Most animals increase surface to volume ratios by folding and packing surfaces into lungs, gills and microvilli. Folding and compacting the folds increase surface area and thereby increase opportunities for transfer without increasing the bulk of the body.

Compartmentalization

At body sizes where simple diffusion is inadequate, systems for compartmentalizing functions develop. No longer is it feasible for food molecules to diffuse through an entire body's surface, so a gut with intake controlled by the head, manages intake of fuel. A kidney, or in our bee's case, her Malphigian tubules, take over excreting wastes, and a gill, lung or tracheal system, exchanges respiratory gases. So, at last, a circulatory system begins transporting substances around the body.

Circulation Combines and Integrates Functions

Bees are efficient. One moving fluid serves multiple functions. Not only does circulating hemolymph transport foodstuffs, hormones, immune cells and wastes, it shuttles heat around the body and dissipates it to the environment cooling thoracic engines that power the wings and legs. For example, heating and cooling are reasons why we might try to emulate insects in building our smallest devices: A flying mosquito maintains less than a one degree centigrade difference between the temperature of its body and the temperature of the air swirling around it. The mosquito maintains this temperature differential despite enormous heat production. Taking size into account, larger honeybees heat up their thoraxes to about fifteen degrees centigrade. Having larger thoraxes and, therefore, smaller surface to volume ratios than mosquitoes, honeybees must generate and then maintain higher temperature gradients before achieving the requisite rate of dissipation of heat. One converse of this principle is that arctic bumblebees are quite hairy. Hair helps retain body heat, but in comparison, tropical bees at low elevations may be quite naked facilitating heat dissipation.

Implications for Devices: Batteries and Reservoirs

Circulations regulate. Physiological mechanisms, in even much smaller insect bodies than those of bees, together with their regulators and suppliers of energy are smaller than half a millimeter cubed. One problem for devices is batteries. Because of their bulk, we must position these energy sources at distances from where we require their energies. Long distances between sources and sinks dissipate useful energy. On the other hand, open circulations draw sources closer to sinks, and reduce excessive energy wastage. Now, what animals employ open circulations, and how do they live?

How Prevalent are Open Circulations?

More animals employ open circulations than closed circulations. This fact alone suggests that the principles of cavity transport are robust. They have served over millions of years through many changes of temperature and climate. Open circulations are important energetically and are here to stay.

History

Open circulations arose prior to Cambrian times perhaps in trilobites when arthropod bodies formed. Some trilobites were slow pelagic swimmers, others faster predators, and some crawled, consuming organic matter from sediments when trapped in the mud of early oceans to deposit their exoskeletons later to be found in Burgess Shale. Among the trilobites, worm like strings of body segments having heads and tails possibly feathery gills occurred (Ref: Trilobites).

Today open circulations, some quite modified, occur in at least three major groups: Arthropods, Echinoderms and Mollusks. Molluskan body cavities are coeloms. Coeloms resemble hemocoels, but they arise differently in development and are highly modified, even though in some mollusks the cavities comprise large portions of their circulation. According to one popular theory, the coelomate theory, mollusks evolved from a coelomate ancestor along with the annelids, the segmented worms, as both show embryonic spiral cleavage, a specific type of cell division during development producing a similar larval form called a trochophore. Such academic questions will probably be clarified after more genetic work. Molluskan cavities will not be considered here.

Arthropods

Arthropods are the largest most successful invertebrate group having now perhaps a million species. Arthropods have exoskeletons containing chitin, and these rigid exoskeletons must shed as their bodies grow. Arthropods are segmented into a head, thorax, and abdomen, but segments may be fused together as are the head and thorax, the cephalothorax, of crawfish and lobsters. Arthropods have jointed appendages. Gas exchange occurs direct through body surfaces in the smaller forms, as well as through gills, tracheae or book lungs that maximize respiratory surfaces. Blood that may or may not transport oxygen circulates within the open circulations of arthropods.

Modern Arthropod groups are three: Chelicerates, Crustaceans and the Uniramia. Chelicerates are spiders, ticks, mites, scorpions, horseshoe crabs and sea spiders. Many of these have book lungs, and most are terrestrial. Crustaceans are mostly marine, but some inhabit freshwater and a few are terrestrial: the fairy shrimps, water fleas, isopods, krill, crabs, shrimps, lobsters, copepods, barnacles as well as a small newly discovered group, the Remipedia. Uniramia include the centipedes, millipedes and insects. Uniramians have one pair of antennae and one or two maxilla, as well as a pair of mandibles. Uniramians breathe through their body surface, gills or tracheae.

Echinoderms (Deuterostomes)

Echinoderms or Deuterostomes are marine and are starfish, sea urchins and sea slugs. Compared to insects, echinoderms are large, cold, slow-moving bottom dwellers. Echinoderms have endoskeletons beneath their skins or epidermis. Endoskeletons are spines or plates having a radial symmetry (starfish), and in some instances, symmetry is bilateral (sea cucumber). Fine networks of branching crystals of calcium carbonate, stereoms, are the building blocks of the interlocking plates or spines. Deuterostomes have an extensive body cavity that for embryonic reasons, is not a hemocoel but a coelom. Like a hemocoel, the coelomic cavity contains fluid and is lined with tissue, so that nutrients and wastes must traverse this interface. In deuterostomes, a water vascular system, a set of water-filled canals, radiates from a ring canal around the gut. The radial canals lead to podia or tube feet on the surface of the body. Tube 'feet' extend and retract according to changes in hydrostatic pressure within the water vascular system. Because deuterostomes are cold and slow-moving, open circulations can supply the low metabolic needs of these animals.

The Circulation of Bees

With the worker bee as our example we explain the major features of a bee's circulation useful for modeling micro-fluidics in engineered systems. We shall not dwell upon the myriad specialized differences between bees and the circulations of other insects.

Hemocoel

Contractions of the dorsal vessel or heart circulate hemolymph within the cavity of the hemocoel. The hemocoel is divided into three sinuses: a dorsal or pericardial sinus surrounding the dorsal vessel, a middle or perivisceral sinus surrounding the gut, and a ventral or perineural sinus above the ventral nerve cord.

Dorsal Diaphragm

The heart lies upon a thin dorsal diaphragm, a sheet stretching across the anterior end of the abdominal cavity in segments three to seven. The membrane holds five pairs of fan-shaped bundles of fine muscle fibers that attach laterally to the anterior ends of the tergal plates of each segment. The fibers spread towards the heart where the fibers break up into many smaller branching fibrils. The pericardial cavity, the space containing the heart, lies above this dorsal diaphragm. The lateral borders of the diaphragm are not attached to the walls of the cavity between the attachment points for the muscles leaving gaps. Hemolymph enters the pericardial

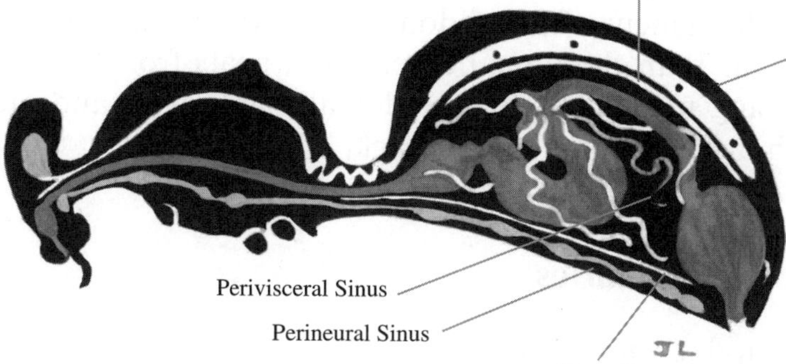

Figure 6.1. The Hemocoel. The hemocoel (black), cut horizontally on the half shell, is divided into three horizontally arrayed intercommunicating compartments: a dorsal pericardial sinus, the space surrounding the dorsal longitudinal vessel or heart, a middle perivisceral sinus surrounding the gut or intestine, and a ventrally placed perineural sinus overlying the ventral nerve cord. A dorsal diaphragm (white line), is a sheet of tissue that is incomplete in many places but separates the pericardial sinus from the perivisceral sinus. In a similar fashion, the ventral diaphragm (white line), also incomplete in places, separates the perivisceral sinus from the perineural or ventral sinus.

cavity through these gaps as they impose little resistance to flow. Muscles as well as the tracheae may occur together in this gap region (see Tracheae below). Sometimes muscles cross the hemocoel from one side of the hemocoel to the other; some muscles even traverse the dorsal diaphragm just below the heart. These muscles contract and compress the abdomen and its contents. Rhythmical movements of the dorsal diaphragm pulsate to ripple the diaphragm in a forward direction and help propel hemolymph forward.

Ventral Diaphragm

A similar ventral diaphragm lies just above the nerve cord. The ventral diaphragm separates the perivisceral sinus around the gut from the perineural sinus around the nerve cord. In the honeybee, the ventral diaphragm extends from the meta-thorax back to the seventh segment of the abdomen. The ventral diaphragm, more muscular than the dorsal diaphragm, beats from front to back, opposing movements of the dorsal diaphragm, and aids hemolymph moving from front to back.

Look Ma, No Capillaries

Remember, the bee's hemocoel does not possess networks of arteries and veins or trees of tubules for distributing materials around the body. In contrast, vertebrate blood in arteries, capillaries and veins stays separated from the extracellular fluid that bathes cells. Unlike a hemocoel, nutrients and wastes in a tubular system must traverse capillary walls, as well as the extracellular fluid surrounding the cells before they pass through cell membranes.

Insect hemolymph bathes and supplies organs more directly than blood in vertebrates, avoiding the added weight, time and the additional energy for molecules to traverse many barriers as they diffuse. As we will see in the chapters to follow, membranes lining the internal walls of the cavity of the hemocoel may act as barriers between cells and the fluid. In some instances, surfaces may be tight, in that hemolymph proper may never directly contact cells, except for the cells floating in the hemolymph and the exposed surfaces of some cells around the heart. However, probably many gaps exist between the cells lining the cavity, and movements of materials across boundaries are complex, highly controlled and precise (Chapter 7).

Hemolymph

The spaces in the hemocoel, unoccupied by organs or tissues, contain hemolymph, that in honeybees is a pale, amber-colored fluid. Hemolymph arises in the embryonic or developing bee as a mixture of 'blood' and 'lymph.' Hemolymph contains many chemicals as well as blood cells or hemocytes that resembling white blood cells of vertebrates. There are no red blood cells in hemolymph. Chemicals include sodium, chloride, amino acids, proteins, hormones, nitrogenous wastes, dissolved carbon dioxide, and many others. The chemical composition and volume of hemolymph change during starvation, desiccation, feeding and at different developmental stages. (For the chemistry and cells of hemolymph and direct measures of the volumes of hemolymph in bees and other insects, see books by Jones, 1997 and Chapman, 1998.)

Functions of Hemolymph

Hemolymph is a reservoir of water and chemicals for the body. Hemolymph supplies digested molecules as nutrients and water from ingested water, honey and nectar as well as ions, hormones, and cells of the immune system to muscles and organs surrounding the hemocoel. Hemolymph transports carbon dioxide to be eliminated through the respiratory organs, gut and cuticle. Hemolymph also distributes heat around the body and aids dissipating metabolic heat to the outside. Hemolymph acquires wastes as the breakdown products of metabolism from the surfaces of cells and organs lining the hemocoel. The excretory organs, the Malphigian tubules, remove these wastes from the hemolymph by filtering and passing them into the gut to be excreted with feces. The water of the hemolymph is about twenty percent of the body water of the bee, but percentages vary. In larvae, the hemolymph may hold up to fifty percent of a bee's body water (Chapter 9).

Regulation of Hemolymph

Hemolymph circulating between the thorax and abdomen cools the motors driving the extremities during flying and walking (Chapter 10). Little data exists on the time taken for hemolymph to make a complete circuit through the hemocoel. Rates obviously depend on the heart rates, activity, developmental stage, degree of hydration and physiological status.

During times of increased and decreased hydration, bees regulate the hydrostatic pressure as well as the volume and chemical composition of the hemolymph to provide a continuous supply of nutrients and energy as well as a stable internal environment. Volumes of hemolymph can rise and fall without risking the circulatory collapse that may occur in mammals and other vertebrates. Good data on the relative amounts of fluid in insects at different stages and under different conditions are rare, and much of this data is quite old.

Molting

Hemolymph forms a reservoir for water and raw materials prior to molting. The volume of hemolymph is large in larvae and when a pupa emerges as an adult or before and during molting in insects that molt. In all insects, some digestion products from old cuticle enter solution to be recycled and reincorporated into the new cuticle. After new cuticle is laid down underneath the old one, epidermal cells and glands secrete a gel-like fluid into the space developing between the old and the new cuticles. When enzymatically activated, this gel digests the old cuticle, and the fluid and its digested products return to the hemocoel before the old cuticle sloughs. One bizarre use of hemolymph is for reflexive bleeding in the Ivory Coast cricket, *Dictyophorus oberthur.* When disturbed, perhaps to protect itself, this cricket covers itself with dense bubbles from the hemolymph.

Volume of Hemolymph

The hemolymph in an insect varies with the species, the diet, the age, the state of hydration and the methods used to measure its volume. Exsanguination, dye dilution methods and C^{14} inulin are the most common ways of measuring the volume of hemolymph.

A bee's volume of hemolymph may be expressed as a percentage of the weight of her body. This percentage of an insect's weight varies with the type of insect and its state of hydration. Generally, hemolymph forms between fifteen to seventy percent of an insect's weight. Consider the hemolymph of an insect as being about 26% of its body weight, a percentage that is greater than the percent of blood volume in vertebrates.

Hemolymph is a reservoir for water. The water in hemolymph is about one-quarter of the bee's total body water. As an example of volumes, one can remove about 0.01 microliters of fluid from a honeybee and about 20,000 microliters from a large queen termite. Some insects appear, however, to possess very little hemolymph in their bodies when attempts are made to extract it from the hemocoel. In general, the smaller the insect, the smaller the volume of hemolymph relative to the size of the hemocoel, but to my knowledge, careful modern comparative studies of this important relationship in large and small insects do not yet exist. One example of a large insect having a small volume of hemolymph is the periodical cicada, *Magicicada* sp. We expect, however, to find low volumes of hemolymph in the smallest forms, as the three-dimensional volume of hemolymph in larger forms shrinks down to become a moistened two-dimensional layer lining the external surfaces of the organs and the inner surfaces of the hemocoel in the smallest insects. A smaller volume of hemolymph reduces weight and should increase the probability of transport within the hemocoel by diffusion (Chapter 9).

Hydrostatic Pressure

Hemolymph, especially in softer bodied forms, such as caterpillars, may work synergistically with the musculature of the body wall as a hydrostatic skeleton. Paralyzing the muscles causes the bodies of soft-bodied insects to become flaccid. For normal tonus, muscles contract against the skeleton of the body wall. As the muscles increase their tone squeezing the hemocoel and its enclosed hemolymph, hydrostatic pressure within the cavity increases by Pascal's principle, in that the increase in hydrostatic pressure transmits undiminished to every part of the volume and to the walls. The pressure within the body depends upon the volume of fluid hemolymph, muscular forces, and resistances to flows of hemolymph within the body. Some insects may even achieve a sub-atmospheric pressure within their hemocoels. Sometimes this pressure is so far below atmospheric pressure that air and fluid may enter an insect following puncture wounds in the cuticle. This poorly understood finding, even occurring when small volumes of hemolymph are present, may have something to do with how rapidly the heart beats. Increased ventilatory movements transporting air into and out of air sacs associated with the tracheae may contribute to these differences in pressure (Ref: Negative Pressures in Hemolymph).

As an insect molts or emerges from a larva or pupa, hemolymph under muscular pressure flows into spaces in the wings and appendages expanding them into their adult forms. Localized regions of increased hydrostatic pressure assist with movements of the body to expand the proboscis. Together with local mechanisms hydrostatic pressure can swiftly autolyse a damaged antenna, leg or wing. In mosquitoes, localized movements of the pharynx and hydrostatic pressure activate a hatching spine of cuticle that pokes a hole in the eggshell to release the larva.

Circulation of Hemolymph

Combined movements of the dorsal and ventral membranes beating together with a pumping heart and pitching and yawing movements of locomotion as well as pulsations of the alimentary canal and other organs together mix and circulate hemolymph.

Peristaltic contractions of the open pump system of the dorsal longitudinal vessel that opens into the hemocoel are the prime movers circulating hemolymph. However, hemolymph in embryonic forms may not appear to circulate rhythmically. Muscular contractions of the body wall, movements of ovaries, air sacs and the alimentary canal as well and walking and flying undoubtedly enhance flow of hemolymph. In some nymphs, for example, transverse septa in the thorax open and close synchronized with respiratory movements, facilitating movement of hemolymph (Jones, 1997: p. 80).

Dorsal Vessel

This single blood vessel, a long slender tube, extends forward along the midline of the back of the abdomen from about the sixth abdominal segment through the thorax and then into the head. In the thorax, the vessel, now called the aorta, loops down between the flight muscles that power the wings. In the head, the aorta, now close to the esophagus, passes under the brain to open into the hemocoel beneath the brain. The abdominal portion of this tube is the heart, and the posterior end of the heart is closed. Even though circulation usually requires a beating heart and assistance from accessory pumps, much hemolymph circulates whenever the skeletal muscles contract. Muscles associated with movements of the pharynx and alimentary canal indirectly propel hemolymph.

Flow of Hemolymph

The aorta directs hemolymph onto the brain. This hemolymph supplies organs in the head before draining backwards through the neck into the thorax and abdomen. Pulsating membranes in the head between the bases of the antennae and elsewhere driven by neighboring muscles force hemolymph to move along well-defined channels that circulate hemolymph through the appendages: the antennae, the wings and the legs. Movements of the ventral and dorsal diaphragms may help to channel these lesser flows. Hemolymph flows from regions of the abdomen around the ventral nerve cord upward into the pericardial space above the dorsal diaphragm. This hemolymph then circulates forward aided by movements of the dorsal diaphragm finally to enter through holes or ostia along the length of the heart. Within the heart, contractions force hemolymph to move again to the head through the aorta.

Pumping

The dorsal vessel, closed at its posterior end, divides into a posterior heart at the back end of the tube and the aorta in front. In this cardiac or heart area, incurrent openings, and sometimes, external openings, the ostia, permit hemolymph in the heart to exchange with hemolymph in the hemocoel. The aorta contains no ostia. Ostia in honeybees occur in five pairs in abdominal segments two to six inclusive. Blood enters the heart through these ostia. Lips of each ostium project forward into the heart cavity. These lips act as valves preventing backward flow of hemolymph each time the heart contracts.

The walls of the dorsal vessel in the heart area contract, as the walls of the vessel consist of sheets of muscle cells wrapped in spiraling layers around the bore of the tube. Systole is the contraction phase. Heart muscles contract synchronously beginning at the rear of the abdomen, and a wave of contraction spreads forward towards the head along the dorsal vessel. The muscles of the heart then relax. This relaxation begins diastole or the filling phase of the heart. The heart fills as it expands and as its muscles relax. Sometimes elastic fibers that pull the walls open when muscular tension subsides assist filling. During diastole, the resting phase of the heart cycle, the heart, now full of hemolymph, pauses briefly. As the heart rate or frequency of contraction increases, the time the heart spends resting in diastole shortens.

The frequency of the heartbeat varies considerably. Frequency of pumping often is higher in early larvae, and the rate of beating may rise just before molting. Contraction of the heart muscles slows below one to five degrees C and stops when above fifty degrees C. Beating may stop for a few seconds and can reverse periodically when waves of contraction begin at the front and move backward. When reversal occurs, hemolymph passes backwards through the incurrent ostia.

Ostia

The incurrent ostia are vertical slits in the sidewalls of the heart. There may be up to twelve pairs of these incurrent ostia in some forms. The front and back edges of each hole are molded into little lips that form valves. These valves allow blood to enter the heart during diastole, the filling phase of the heart. During diastole incoming blood forces the lips apart, and hemolymph flows into the heart from the hemocoel. The valves prevent backflow through the ostia when the heart contracts and pumps during systole. Systole is the ejection phase of the heart. The pressure of hemolymph in the heart squeezes the lips of the incurrent ostia together as the heart contracts and the lips remain shut during systole. In some insects the heartbeat reverses, and when reversal of flow occurs as the heartbeat reverses, hemolymph can flow backwards into the hemocoel as the heart contracts.

In the groups where the excurrent ostia occur, little clumps of cells, called papillae, sit at the entrances of the ostia into the heart. These papillae expand during systole forcing hemolymph out of the heart into the hemocoel. During diastole, the papillae contract preventing hemolymph from flowing from the hemocoel back into the heart.

Control of the Heartbeat

In many groups of insects cardiac nerves from the nervous system control the beat of the heart. In others groups, the heart is myogenic, and the heart receives its stimuli to contract from pacemakers in the heart muscles themselves. Hormones may also modulate the heartbeat. Hormones may arise from the terminations of neurosecretory nerves that end directly upon heart muscle, or hormones may enter the hemolymph in the hemocoel from organs distant from the heart.

Accessory Pulsatile Organs

Accessory pumps associated with the wings, legs and antennae move hemolymph into the extremities. In some insects, the pulsatile organs of the wings may be derived from the heart pump directly, but in bees these pulsatile organs are separate from the heart. Accessory pulsatile organs maintain circulation within the appendages. Flow to and from the accessory pumps mixes hemolymph moving in the hemocoel with hemolymph circulating through the appendages. Contractions of these accessory pumps may be intrinsic to the muscles that comprise them, or nerves may control the pumping.

Extremities Divided

Extremities such as legs or antennae each may have a membrane like a diaphragm down their centers. These membranes often separate flexor muscle groups from groups of extensor muscles on the opposite side. When muscles on one side of a leg, the flexors let's say, contract, to raise a leg, the contracting muscles squeeze hemolymph from this flexor compartment back into the hemocoel of the thorax. When the insect steps down, muscles in the opposite compartment, the extensor compartment, contract and squeeze hemolymph on their side of the membrane back into the thoracic hemocoel. The relaxing compartment alternates in each phase of stepping and filling with fresh hemolymph from the hemocoel each time. There are several types of pumping mechanics for the accessory pumps in different groups (Chapman, 1998).

Reviewing the Path

During normal circulation in the worker bee, the heart pumps hemolymph forward into the aorta during systole. The blood then enters the perivisceral sinus in the head as the aorta opens out over the brain. At this time, blood also exits the dorsal vessel through the excurrent ostia. Valves within the incurrent ostia close during systole, so that hemolymph does not leave the dorsal vessel as the heart squeezes. Were the incurrent ostia in the heart to open during systole, pressure would be lost, and less hemolymph would pass through the aorta.

As hemolymph leaves the aorta to enter the perivisceral sinus, the new hemolymph entering the sinus forces the hemolymph from the last systole backwards through the sinus. Movements of the dorsal diaphragm aid this flow. Slowly moving hemolymph washes over the organs and tissues where exchange occurs. As the heart pump fills during diastole, hemolymph leaves the perivisceral sinus in the abdomen to enter the heart. As the heart sucks hemolymph into itself during filling, loss of hemolymph from the hemocoel promotes rearward flow of blood through the perivisceral sinus. Presumably, a moving perineural or ventral diaphragm encourages blood to flow around the ventral nerve cord supplying it.

Thoraco-abdominal Shunting

In some active insects, butterflies, beetles and flies, hemolymph shunts back and forth between the thorax and abdomen as a flap of fatty tissue or large air-filled sacs reverse periodically near the pedicle when pressures in the abdomen and thorax shift back and forth, alternately opening and closing a tubular connector in the pedicle between the thorax and abdomen. When the abdomen contracts, abdominal pressure rises to exceed hydrostatic pressure in the thorax so it pumps hemolymph forward into the thorax and head. When the heart reverses and pumps hemolymph backwards, this new volume actively expands the abdomen, and the expansion draws hemolymph past the flap of fatty tissue. In such active insects, movements of the hemolymph correlate with movements of the tracheae. These reversals of pressures and flows serve to control oxygen flow to the muscles of the wings and legs in the thorax and to exchange heat from these same muscles through the abdomen (Chapter 10).

Some insects circulate hemolymph through their wings. This circulation may occur in young adults as wings stiffen and become flight-ready. In the absence of circulation through the wings, the tracheae in the wings collapse and the wings turn dry and brittle. In forms having hemolymph moving in the wings, the thoracic pulsatile organs control this flow.

Respiration Compared

The vertebrate circulation lies between the lungs or gills transferring oxygen and the cells using it. In insects, the respiratory system is separate from the circulation. Were a vertebrate-like system used in insects, insect size would disadvantageously need to increase. The presence of a respiratory tree within the chain of respiratory gas exchange may have imposed minimum size requirements on vertebrates that substantially exceed the needs of insects.

Energetics

Flying consumes huge quantities of energy. Metabolism of flying insects depends upon an extremely efficient and miniaturized system specialized to deliver oxygen to where it is used and to remove heat and carbon dioxide from where these are made.

Thoracic flight muscles have the highest known mass-specific rates of oxygen consumption. Tracheoles are oxygen's closest contact with tissues. Blind-ended tracheoles often penetrate deep into muscles reducing the distances oxygen must travel. Diameters of tracheoles are about the distance of the mean free path-length for oxygen. The mean free distance is the distance a molecule travels on the average before it collides with another molecule. Oxygen enters the mitochondria, the cellular powerhouses, that supply the phosphate containing ATP molecules the contracting muscles need for aerobic metabolism. Mitochondria oxidize fuels they receive from the hemolymph to make carbon dioxide and water. Molecules of carbon dioxide exit the mitochondria of the flight muscles, to circulating hemolymph that transports them through the body and delivers them to the air.

Tapering of the abdomen compensates the limits to diffusion of heat and carbon dioxide from respiration by the exoskeleton and the cuticle and the nature of the tracheae. Bernoulli entrapment of air in the tracheae aids gas exchange. High external velocities of airflow over the posterior spiracles during flight create a gradient of pressure. This gradient drags air into the anterior spiracles and, thereby increases bulk flow of air through the respiratory system. Respiration and circulation are coordinated in that respiratory and locomotory movements also help to circulate hemolymph.

Graph of the Hemocoel Resembles an Almost Random Graph

How do the dynamics of a hemocoel depend upon the structure of its graph? Graphs can help us understand how the connectivity of a hemocoel functions, as graphs can generalize the connectivity of lattices and trees. The only way to move through a lattice or a tree is to crawl from vertex to vertex. There is no action at a distance, nor are there secret wormholes magically transporting a crawler from one branch tip to another universe. Graphs may show us shortcuts permitting more flexible arrangements or new ways of traveling and do not always rely upon regular step-by-step connections. Graphs let us violate the principle of locality.

The graph of the hemocoel approaches that of an almost random graph. To build a graph of a hemocoel, begin with a collection of *n* vertices and no edges. The vertices would be the places transported molecules might enter or leave the hemolymph by crossing a membrane. Circulating hemolymph allows pairing of any one vertex to any of the others. We create edges with probability *p* if a molecule or particle makes or can make the transition from one vertex to another vertex through the hemolymph, but we have an edge being created with a probability of zero if the transition does not occur for any reason.

In the extreme case of a hemocoel empty of hemolymph, our graph remains edgeless as no transitions are possible, but if the hemocoel is filled with hemolymph, the graph becomes a clique. A clique graph has all pairs of vertices adjacent. Between these two extremes at intermediate volumes of hemolymph, we might expect the graph to have all edges placed randomly and independent of each other.

Consider a complete graph, K_N, having N vertices or nodes and $N(N-1)/2$ undirected edges. To each edge we attach an exponential random variable, Eij having a mean equal to 1. For any two arbitrary nodes we want to know the random number of edges in the shortest path connecting these two nodes. As the hemocoel fills, the size of its graph goes to infinity, so the probability of an edge exiting between two vertices now grows larger than some threshold value, so that the graph of the hemocoel becomes connected. The strength of the random graph of the hemocoel is that the graph can be as dense or sparse as necessary. This change occurs by adding or subtracting edge probability, *p*. Because each new edge forms or disappears independently from any other, the graph does not form clusters, because neighboring vertices are now no more likely to be linked than would be any other randomly chosen vertices. Unlike a closed tubular circulation, nodes are eliminated in the graph of a hemocoel showing the hemocoel system to be less vulnerable to point defects (Ref: Random Graphs).

Changing Volumes and Shortest Paths

As the volume of hemolymph in the hemocoel rises and falls, the shortest paths between the most distant vertices, varies. Newman, Watts and Strogatz (2001) studied the minimum path length, L, averaged over all pairs of vertices in graphs to find an abrupt transition in L as the 'rewiring' probability increased. L, maximal in a regular lattice, fell steeply when they rewired just a few of the edges. Watts and Strogatz defined a clustering coefficient, C. To calculate C, list all the neighbors of a vertex, count the edges linking those neighbors, and then divide this number by the maximum number of edges that can exist among the neighboring vertices. The operation repeats for all vertices, and the average is taken. In contrast to the path length L, the clustering coefficient remains large until the rewiring probability is quite high, so that over a wide range of p values, local connections between nearby vertices dominate the graph, but a few shortcuts provide for the necessary long-range connections.

How Do Molecules Know?

Shortcut connections between vertices are of little use if a particle in the hemocoel does not know where to go. How does a particle, lacking a nervous system and a topological map of the hemocoel in its brain know the most direct route?

We can modify a two-dimensional square lattice where each vertex joins its four nearest neighbors. To this lattice we add long distance connections, but not merely at random. We rank each edge connecting two vertices to all possible destinations as we would a shortcut edge. We rank each edge based on its distance from the source vertex. Now, the probability of choosing a vertex,

d, is proportional to d^r where r is an additional parameter of the model. If we set r equal to zero, then we choose destinations at all distances from d with equal probability, so our model just becomes a two-dimensional version of Watts and Strogatz's model. However, if r is large, then our omniscient molecule has appreciable chances of choosing only nearby destinations, so the structure of our starting lattice barely changes. The threshold value is when $r = 2$. At this point, the probability now obeys an inverse square law (Ref: Routing).

A greedy algorithm can solve such routings. To go from vertex a to vertex b, list all the edges emanating from a. Now choose the one edge taking us closest to b as a measured distance across the lattice. Then repeat the same procedure starting from our new vertex, again proceeding vertex by vertex, until we reach our destination.

Degree Sequence

Degree sequence is simple in a lattice. All vertices have the same number of edges, so a plot of the degree sequence reveals a single sharp spike. Any randomness in this graph broadens the peak into a Poisson distribution. Because of the exponential decline, the probability of finding a vertex with k edges grows negligibly small for large k.

Does a power law describe the edges in the hemocoel? That is, are the number of vertices having degree k given not by e^{-k} but by k^g, where g is a positive constant? The power law distribution falls off more gradually than an exponential. This gradual decline permits vertices of high degree to exist in the hemocoel.

We can use Barabasi's model of edges in the World Wide Web to study the hemocoel. Begin with n vertices and no edges. At each step, add to the graph a single new vertex and m edges, so that now all new edges link the new vertex to some of the vertices already present. This pattern fits with what happens as volumes of hemolymph in the hemocoel vary as edges disappear from or add to the network depending upon the volume of hemolymph in the cavity.

The probability that a given vertex receives a new edge is proportional to the share of the total set of edges that the vertex already possesses. Hence, a well-connected vertex upon filling of the hemocoel, evolves into a more connected vertex. After t steps, the graph has $n + t$ vertices and m times t edges. Growing by Barbasi's rules, the graph enters a statistical steady state, in which the shape of the distribution of vertex degrees does not alter over time provided the levels of hemolymph remain constant. The probability of finding a vertex having k edges is proportional to k^{-3}.

At this point, the graph of the hemocoel approaches that of a Moore or random graph. In a Moore graph starting from any vertex, we can reach k vertices at distance 1, then from each of these vertices we can achieve another $(k - 1)$ new vertices at distance 2 and so on without any redundancies until we have filled the hemocoel with its graph. A Moore graph is the most efficient possible k regular graph in the sense that every vertex "reaches" k new vertices. For a perfect Moore graph, a theoretical lower bound that we seldom reach is that for $k > 2$, the characteristic path length has to grow at least logarithmically with n.

Connectivity Reviewed

We consider all points in a hemocoel, the vertices or nodes, to be identical, un-weighted and featureless, so connections of one node with another through the hemolymph become our graph's edges. The degree of a vertex is the number of other vertices with which it connects. We consider only the transmission possibilities through the hemolymph as forming the network. For example, transmission of a particle from one side of an organ to its other side within the organ, and, therefore, outside the hemocoel, does not constitute travel within the hemocoel. Travel through the hemolymph from one vertex to another is bidirectional as transmission is equally likely in both directions. The graph of the hemocoel is sparse. Φ measures the number of shortcuts or the average range if the graph is large, so that the hemocoel graph exhibits logarithmic length scaling with respect to n, its number of vertices. One might presume a measure of Φ should be close to the characteristic path length for the graph in that edges do not correlate with each other (they are merely paths through fluid in two or three dimensions). Because two vertices are connected does not imply that their second shortest path length should be any shorter than the average path length. Gamma, indicating an undirected graph, and L, indicating a line graph, are larger than for a random graph. Gamma refers to length, so clustering approximations fall apart because any Moore graph approximation of local scale diverges.

Functions of the Volume of Hemolymph

Hemocoels remain the same size, but the volume of hemolymph, and hence the network graphs inside the hemocoels, grow and shrink. Hence, the degrees of the vertices cannot remain constant. At certain sub-maximal volumes of hemolymph, the vertices will have skewed the degrees of connectivity. As a consequence of these volume dependant changes and others over time, the connectivity of the graph of a hemocoel will vary from Moore type to some related modification of the random graph. If *pk* is the probability that a randomly chosen vertex has K neighbors, it turns out that *pk* has either a power law tail as a function of k, indicating that there is no characteristic scale for the degree or a power law tail truncated by an exponential cut off. Such distributions differ from the single scale Poisson distribution in traditional random graph models of networks (Ref: Graph Models of Networks).

Open Circulation: Advantages

Unlike in a closed circulation, no hubs exist in hemocoels. Connections within the hemocoel are direct and bidirectional, and because every point within the hemocoel is potentially alike and potentially connects with any other, hemocoels are not vulnerable to point blockages. We assume all routes for transmission within hemolymph to be identical, although transmission may not always equal physical flow. One example might be a cell creeping over the surface of an organ. As an exponential network, circulation inside the cavity is more stable than a pure scale free network would be. As we shall see, circulation can grow relatively independent of the size of a hemocoel and its volume of hemolymph, and shortcuts may be of indeterminate length (Chapter 10).

Review of Hemolymph

Hemolymph receives nutrients from the gut and water and ions from the rectum in the last part of the alimentary canal. Hemolymph supplies water to the organs and tissues, and water consumed in honey and nectar can enter the hemolymph through the alimentary canal. Hemolymph is a reservoir for water for the body. The water of hemolymph holds about twenty percent of the body water of the bee, but its percentage varies. In larvae, hemolymph may contain up to fifty percent of the larva's body water. Hemolymph distributes nutrients, ions and hormones to muscles and organs surrounding the hemocoel and transports cells of the immune system to where they are needed. Numerous hemocytes or blood cells float in the hemolymph. These cells resemble the white blood cells of vertebrates and are of several kinds. Hemolymph distributes digested molecules that are absorbed from the alimentary canal. Hemolymph receives the breakdown products of metabolism that are later removed by the excretory organs, the Malphigian tubules. Hemolymph also transports carbon dioxide to be eliminated through the respiratory system and cuticle and as bicarbonate with the feces.

Regulation of the volume and compositon of hemolymph as well as changes in its hydrostatic pressure occur during times of increased and decreased hydration, to provide a stable cellular environment and a continuous supply of nutrients and energy to active areas and muscles of the body. Volumes of hemolymph can rise and fall without endangering circulation that can occur in mammals and other vertebrates. Good data on the relative amounts of fluid in insects at different stages and under different conditions are rare, and much of what we have is quite old.

Chapter seven

WHERE THE HEMOLYMPH MEETS THE WALL

Overview

Where the hemolymph meets the wall is where the rubber meets the road. Here we consider the walls and surfaces that hemolymph contacts. Walls are the portals to and from the moving liquid medium. Activities in or on the walls determine what enters and leaves the hemolymph and, hence, determines distribution to and from the body.

Walls include all surfaces contacting hemolymph: the external surfaces of organs such as the coverings of the alimentary tract, organs and muscles, the walls of the hemocoel itself, and the diaphragms. The diaphragms help direct the bulk flow of hemolymph. Central regions of the diaphragms direct flows fore and aft, but flows above and below a diaphragm may join along their lateral edges.

The circulating bulk volume of hemolymph transports molecules. Substances then diffuse from this bulk flow into surface layers of fluid along the walls before being absorbed. The local geometry and mobility of the surfaces and any microconfigurations abutting their interfaces help or hinder adsorbtion and desorbtion.

Volumes of hemolymph, pumping rates, and walking and flying determine how the hemolymph moves and circulates. Surface geometries and localized patterns of flow determine which molecules adhere to what points as well as the rates of reactions and release. Surface functions entail many highly irregular complex interactions that continuously evolve over the life of a bee or device.

To point up the similarities between bees and devices, consider now a hemocoel as if it were a microfluidic device.

Hemocoel as a Microfluidic Device

Microfluidic devices contain small volumes of fluid and reagents; they are inexpensive to make, and devices having diverse functions can be mass-produced in groups or individually to perform in concert on a single chip. Applications for a hemocoel-like device should be myriad.

Generalized Microfluidic Chip

A microfluidic chip is a plate of silica, glass or plastic trenched with narrow channels through which samples flow in tiny streams. Blood, bacterial suspensions, and solutions of proteins and buffers flow within the channels. Architectures of the channels and chips vary. Channels may be open or closed, and the walls are coated with substances that extract molecules from solution. Only molecules touching the walls can be extracted, and most fluid and solute flows unimpeded along the centers of the channels. Currently, devices measure diffusion coefficients, viscosities, pH, reaction kinetics and other modalities. Molecules cross walls bidirectionally as hemolymph gives up or receives molecules for cells residing on, beneath or in the surfaces. Both chips and hemocoels rely on surface-to-volume ratios for extractions.

Principles of Flow

Reynolds number, *Re*, characterizes flow. (Note at this chapter's end.) For many devices, D, the characteristic length, equals $4A/P$, A being the cross-sectional area of the channel and P the wetted perimeter of the channel. In devices, *Re* is much less than one hundred and often less than ten, so flow remains laminar and non-turbulent. Laminar flows transport molecules predictably, but in bees at least, changes in the momentum of hemolymph probably alter these flows.

Pressure driven syringe pumps drive flows through the channels of microfluidic devices so that single layers of fluid contact the walls and remain stationary. Slow flow over the walls creates a parabolic profile of fluid velocities across the channel. The most centrally positioned flows encounter the least resistance and, therefore, travel the fastest. A second method for driving fluid is electro-osmotic pumping (EOP).

EOP distributes a double layer of oppositely charged ions along the walls of the channel. An electric field positioned across the channel encourages ions in this double layer to move toward their opposite polarities. Viscous forces impede and resist the ions moving along the walls. Convection drags the bulk of fluid in the center of the channel. Electro-kinetic flows require high, superimposed voltages and will not be considered.

Models

Models are essential aids for designing devices. Computerized simulations generate models having varying channel geometries, flow rates, diffusion coefficients and chemical interactions. Simulations coordinate these into larger numerical models. For the simplest systems, models work almost as well as mock-ups. Industry creates examples of these models almost daily, so rather than give one here which will be soon out of date, use a search engine to find the latest.

Equations derived from principles of mass, momentum and energy that govern flows in microfluidic devices include the complete Navier-Stokes continuity equations as well as several modifications of these equations that spatially discretize differential equations over a solution domain, such as the finite difference and finite volume methods and the finite element method (Ref: Flow Equations).

H-Filters

H-Filters function analogously to hemocoels. H-Filters continuously extract specific molecules from mixed solutions that may contain interfering particles (cells, debris). Like hemocoels, H-filters lack membranous filters that must be cleaned periodically. Because the Reynolds numbers of H-filters are less than one, and flows are slow, convective mixing does not occur. Diffusion alone transports molecules transversely across the channel of an H-filter. The root mean square distance a molecule travels in an interval is the square root of $2Dt$ (the Einstein equation) where D is the molecular diffusion coefficient. D scales roughly with a molecule's size, so if we ignore charge, small molecules flow faster than larger ones. The time to traverse a channel within an H-filter is proportional the channel's length, so molecules having different diffusion coefficients separate themselves along the channel. As with a five-yard dash compared with a mile race, however, slow runners distinguish themselves from faster runners only when the course is long enough to permit their separation.

Workable reproducible assays utilize the profiles of velocity for molecules in channels having different aspect ratios. Higher aspect ratios permit quantitative studies. Because the relative velocities of two fluids moving in a channel can determine the width of the stream flowing across the width of an H-filter, we can use this idea to create 3-D models. If two fluids have the same viscosity, each occupies half the channel. But if a fluid is paired with a fluid of a higher viscosity, the fluid of higher viscosity flows slower and comes to occupy a greater portion of the channel (Ref: Yagerfaculty).

Micro-rheology

Micro-rheological studies reveal properties of fluids at the micron scale. One seeds a flow with small particles (typically a few hundred nanometers in diameter) and observes their motions under light microscopy. Brownian motion jiggles the particles, but particle-tracking algorithms can follow these motions if we illuminate the particles so they form bright spots in videos. Algorithms reveal the visco-elastic nature of the suspending fluids. For example, the mean square displacement of particles in a Newtonian fluid is simply proportional to a particle's molecular viscosity (via the Einstein equation), but more complex non-Newtonian fluids, polymers and protein-laden solutions in hemolymph, for example, behave differently.

The microstructure we can see under light microscopy and the composition or weight fractions of elements in the mixture determine the physical properties of traditional solutes. For nanostructured materials, however, the atomic species present, and their configurations on the atomic scale determine the properties. Surface roughness determines the wetting properties of liquids at both the micro- and nano-scales. Exact molecular shapes rubbing over each other determine boundary conditions and wall slip. Newtonian fluids encountering highly hydrophobic surfaces may slip. A rough surface at molecular dimensions may inhibit slippage. Fabrication technology permits constructing atomically smooth surfaces as well as precisely setting the heights of the micro-channels.

We can model flows in devices and hemocoels having two and three dimensions. In two-dimensional models, we give the channels infinite dimensions and therefore no top or bottom wall boundary conditions. Such two-dimensional views resemble our concept of a hemocoel as a single surface enveloping the hemocoel that forms when fluid volumes are very small in the smallest insects.

Minimal Surfaces are Stable

Nature favors minimal surfaces because minimal surfaces are physically stable. Minimum surfaces are common: bubbles, planets, and cells to name several. The eardrum spanning the handle of the malleus and the annulus and the membranes between some cells are minimal surfaces. Having a minimal area of surface means that the surface stores minimal energy. Because surface energy is directly proportional to a surface area, as a result of minimizing the surface energy, the area of the film is least when compared to the areas of neighboring surfaces that might span any given contour. Surface-active materials such as detergents can alter surface energies. Much of a hemocoel's surface may approach minimality, especially as a hemocoel and its contained surfaces shrink. Interactions occurring on, against or within a minimal surface will influence the character and rate of activities on this surface.

To visualize a minimal surface, dip a circular wire into a soapy solution to obtain a soap film spanning the loop. The loop holds the flat surface open. This surface is minimal, because of all the surfaces that might span the loop this one film possesses the least area. If you now dip two circular wire loops and hold one loop a short distance above the other, a film spanning both wires, a catenoid, arises. No surface bounding both wire loops has a smaller area. The curvature of the catenoid in one direction is equal and opposite to its curvature in the other, so the mean curvature of the catenoid is zero. Therefore, the catenoid, like the plane, has no curvature and exerts no pressure. So now we may imagine that both these minimal surfaces might span a hemocoel allowing the catenoid and the plane to co-exist as parts of one and the same system of boundaries in the same way two hemispheres can cap the open ends of a cylinder. A catenoid surface often bounds the cooling towers of power plants.

Minimum Surfaces Imply Minimum Weight

Minimal surfaces within a hemocoel can minimize overall surface tension within the hemocoel. Having most surfaces minimal surfaces minimizes stress on the structures that support them. Energetically it is well that weight and thickness of the supports for the hemocoel be minimized. Insect exoskeletal supports are not heavy and bulky not only to conserve space. The energy to hold the hemocoel open needs be minimized as well. In dissecting fresh specimens of many insects, the walls and surfaces of their hemocoels appear to be subjected to very low wall tensions (personal observation). Low wall tensions tend not to implode the exoskeleton, thereby permitting exoskeketal buttressing and cuticles to be as thin as possible. Hence, theoretically at least, we may view the walls of hemocoels to resemble soap-films and the insect system to be minimized along these lines in many ways.

Contours of Co-axial Circles

As with our soap film catenoid covering two co-axial wire loops of the same radius lying parallel to each other, we can visualize a hemocoel as providing a contour for more than one film. How many films or how much surface area might we span within a hemocoel's contour? Schoen showed that we are able to span only minimal surfaces of revolution on such contours (Ref: Minimal Surfaces).

Consider an ideal shrinking hemocoel. Hold three coaxial circles parallel to each other, one of which lies in plane $z = 0$ and the others in planes $z = +1$ and $z = -1$. Using examples of Morgan and Gulliver-Hildebrandt we can generate contours that remain invariant under rotation. We may also add several segments together such that the generating system can now be split into two symmetrical parts that no longer are invariant under rotation to give us a minimal surface that spans the contour of our original hemocoel. Because our second catenoid is unstable, by using soap films we obtain just one of the two possible catenoids: the one that ends up being most like a cylinder (Formenko and Tuzhilin, 1991).

Deformations of Interfaces

We must keep in mind that a surface, apparently smooth at low resolution, may reveal complicated structure under higher resolutions. At all spatial scales surfaces are like coastlines; surfaces appear bumpy at different scales. Objects whose magnified pieces are similar or look similar to the whole are self-similar. The lengths we measure depend on the lengths of our rulers. Scaling means that the properties of what's observed depend on the scale we use to measure them.

We can show how the area of any surface in three dimensions may increase if we dimple the surface so that no surface can become a local maximum for its area functional. Critical points other than a dimpled maximum or minimum are saddle points of the area functional. Minimal surfaces at saddle points are unstable, so even small fluctuations in amplitude may lead to collapse. Think of blowing on a soap bubble. The bubble indents at one location but bulges out at another, but the volume of the bubble remains unchanged unless our breath warms the bubble. Now, given a hemocoel at its smallest dimensions in which the hemcoel behaves as a two-dimensional sheet, assume its infinite planar surface to be minimal. Now a closed curve anywhere on this minimal surface will have the smallest possible area having the curve as its boundary.

Before continuing, we must keep in mind that the mathematics of surfaces suggest simple physical "solutions" that nature never realizes completely, because real surfaces and fluids are never entirely "homogeneous." Asymmetrical forces and chaos always intervene to perturb surfaces and systems that approach equilibrium (Meakin, 1998).

Surface Effects

Flow patterns looping in and around conformations in a surface are complex. Uptake and discharge mechanisms involve transport across surfaces interfacing between two media.

Might we infer from considering the surfaces of the hemocoel what events might transpire on or near them as molecules of nutrients, wastes and hormones traverse the boundaries outward from cells to hemolymph and inward from hemolymph into cells?

Analogous surfaces include digestive and respiratory surfaces. We know from studies at the microscale level of resolution that respiratory membranes and intestinal surfaces are complex and irregular. Both have in common extended surface areas that enhance trans-surface transport dynamics. Such extended surfaces are often folded, packed or rolled up conserving space. We minimize the volume subsumed by a surface if we fold, crumple or compress larger surfaces. Even if such surfaces behave within limits as ergodic fractals, randomness presents similar patterns to surfaces existing across different scales.

Over the course of a lifetime, surface structures may evolve through developmental sequences passing through an array of forms and functions. We can estimate fractal dimension from the slope of a log-log regression curve, but to do so we must discard much structural information. For example, as we have seen, insects have solved distribution problems uniquely by coupling a direct point-to-point distribution system with a more general one. The tubular system of tracheae and tracheoles supplies gaseous oxygen diffusing through air directly from the spiracles to metabolizing muscle while simultaneously, foodstuffs and wastes distribute generally to these same tissues through the circulating hemolymph.

What happens when hemolymph encounters surfaces and contours as it percolates from the aorta back through the cavity of the hemocoel? By extension, how might technology duplicate what happens in the hemocoel-hemolymph system in microfluidic chips or other systems composed of materials such as porous polymers?

Service Areas and Surfactants

Mechanical and chemical properties of surfaces often interact to determine how liquids pass over or near surfaces. What adheres to an interface depends on how it adsorbs as well as upon the surface's elasticity and viscosity. Many parameters feed back controlling how dispersed and dilute or concentrated the bulk of the hemolymph becomes. Hemolymph composition fluctuates as matter passes from the hemolymph into the cells through surfaces and as new materials enter the hemolymph from the surfaces. However, such changes have not yet been studied. Curving of a surface creates localized regions of differing pressures in the hemolymph. Finally, capillary effects themselves may affect the global hydrodynamics for transfer, as the volumes of hemolymph and the wetting of surfaces change with an insect's hydration.

Surfactants stabilize the moving colloidal systems of films, drops, bubbles and foams. Stabilization alters systemic properties and can influence how devices work. Can learning about insect surfactants teach us how to improve our own? After all, volumes of insect surfactant must be quite small, so insect surfactants undoubtedly are very capable of influencing boundary conditions in extremely small systems that could have great practical importance for us. Surfactants also resist clotting in the bulk flow but may encourage clotting if hemolymph exudes from a cut in the cuticle. Hysteresis probably changes as a hemocoel changes shape. Hysteresis effects are common in lungs that change size on inspiration and expiration (Ref: Hysteresis).

Smooth or convoluted, surfaces of the hemocoel efficiently contact the hemolymph flowing over them. Molecules can be thought of as following a concentration gradient driving them in their directions of movement. When an organ accepts molecules through its walls, the molecule's concentration in the hemolymph decreases, unless of course, its supply from the midgut or other source keeps up with removal matching demand.

Hemocoel as Pump

We might even conceive of the whole hemocoel functioning as a low-pressure pump for hemolymph. Studies on the Echiuroid worm *Urechis caupo* and the polychaete *Hermothoë imbricata* suggest the cavity of the hemocoel may be analogous to a spring, and even the hemocoels of caterpillars and perhaps bees may behave similarly (personal observations and Ref: Lawry).

Muscular contractions and relaxations in the wall evoke changes in the internal hydrostatic pressure of the hemolymph. Proportional changes in volume occur as muscles relax and the wall expands. In an extreme case, consider a caterpillar feeding on and digesting leaves so that liquid enters the hemocoel through the midgut accumulating in the hemolymph already distending the walls of the cavity. We might expect if we plotted volume change against pressure change, that we'd get a straight line, but as in filling and deflating vertebrate lungs, the filling curve does not necessarily follow the emptying curve. As the volume in the hemocoel increases, the loop of the emptying curve widens so it is not the same as the filling curve. What is happening? For any volume, filling pressure on the emptying curve is less than this pressure on the filling curve. Therefore, elastic recoil of the walls during the emptying phase is less. The distending transmural pressure gradient had to be higher at each volume to inflate the cavity. This manifestation of a loss of energy as the hemocoel recoils that obeys Hook's law is hysteresis.

If we slowly inflate a caterpillar's hemocoel and leave it inflated, in about a minute the pressure drops exponentially to around fifty percent of the initial value just after filling (personal observation). As in many systems, the hemocoel remembers its recent history in that low volumes of hemolymph sustained over time soon are followed by a reduction in compliance. Hysteresis may occur from stress relaxation of the materials of the wall and surfaces, redistribution of hemolymph within the cavity to areas having differing time constants, changes in surfactant activity with volume changes, as well as changes in the absolute volume of hemolymph. We must remember, however, that these ideas have not been carefully studied, and that in the lung at least, total compliance of the system as a whole probably bears little relation to the combination of compliances in the surfaces of the hemocoel.

Surface Tension

A surface is a notion of geometry. For surfaces whose radii of curvature approach molecular dimensions these concepts become ambiguous. A real interface is a three-dimensional non-uniform region interposed between two bulk phases. Gibbs and others employed a geometrical surface they imagined as a dividing surface interposed between the two phases.

Rigorous definitions of the interfaces employ metric tensor fields and a system of orthogonal curvilinear coordinates that reduce the metric tensor to its diagonal form at any point. We locate our surface between the two phases in relation to the mean positions of the molecules in question after we statistically average these molecules over their disordered thermal motions or in terms of the distance of closest approach of the molecules of one phase to those of the condensed phase. If we place our dividing surface appropriately, we may choose adsorption equal to zero. The surface amount or Gibbs adsorption may be positive or negative, and we define this adsorption to be the excess of the amount of a component molecule in our system compared with that present in a reference system having the same volume as our system in which the bulk concentrations in the two phases remain uniform up to the Gibbs dividing surface.

Adsorption of a component of a multiphase multi-component system such as hemolymph occurs if concentrations of the components in the interfacial layers differ from those in the adjacent bulk phases. The Gibbs dividing surface is a geometrical surface chosen parallel to the interface. If the hemolymph contacts a wall, the boundary surface becomes this dividing surface. Two surfaces can meet along a linear interface. We can also have foam as an array of films and channels and regard these in our models as surfaces and lines. Line tensions of the channels play roles in forming foams (Ref: Capillary Hydrodynamics).

Bulk Hemolymph: A Solution of Surfactants

Surface properties relate to each other as well as to the bulk properties of the hemolymph. Surfactants form an adsorbed monolayer. We might imagine molecules as parking themselves within this adsorbed monolayer. This model works well for non-ionic surfactants. If the surfactant within the volume of hemolymph is dilute, surfactant in the monolayer per unit area coincides with the adsorption of surfactant. We may define the idea of limited parking areas that limit adsorptive capacity as being the monolayer capacity.

The monolayer of surfactant exerts a two-dimensional pressure over the surfaces. Films are fluids spread more thickly than as monolayers, in that a film contains a bulk layer interposed between its two surfaces. In accord with the Gibbs theory of capillarity, we can describe film like interfaces if we replace surface tension with film tension.

Adsorption and Desorption

Adsorption not only refers to attachment but to bulk transfer of a substance from the bulk of the hemolymph onto the surface. To create an adsorbed monolayer, molecules of surfactant may move from the bulk of the hemolymph to the surface, or the surface, as in the lung, may produce a surfactant locally. Molecules then enter the monolayer to pack or park themselves and orient themselves. The first stage is diffusion; the second stage is 'pure adsorption.' Relative roles of these two stages may differ for different surfactants and molecules, thus determining the mechanism of adsorption and perhaps in cases of the thinnest surfaces, the stiffness of the surface as a whole.

Transfer To and Across Irregular Membranes

Transport across the wall of a hemocoel is analogous to transport across rough or porous surfaces in batteries in which high currents flow from porous electrodes. Fractal surface irregularities enhance both processes. Diffusing particles emitted by a diffusion source walk randomly. If a random walker collides with the electrode or membrane, it is absorbed with a finite probability, termed the sticking probability. The sticking probability corresponds to a finite permeability of the membrane. In diffusion, most particles diffuse back towards the source, because the net flux due to Fick's law is proportional to the gradient of concentration and not to the concentration itself. Net transfer by diffusion between the hemocoel and the surface is due to the few molecules that absorb onto the surface before they can return to their bulk source in the hemolymph. Hence, there are two limits to the efficiency of transfer. First a molecule must reach the surface; second it must enter it. This idea is analogous to ions being transported through an electrolyte and then undergoing a redox reaction on the active electrode of the battery to produce the current.

Deformation of Fluid Interfaces

Changes involve fluctuations in surface area as well as changes in adsorptive and deadsorptive fluxes. Rigorous solutions of such problems require knowledge of the non-steady regions of flow by dividing the volume into regions of constant bulk concentration surrounded by thin diffusing layers at the interfaces where the concentration of surfactants changes linearly.

Porous Monolithic Polymers

Porous monolithic polymers, developed over the last ten years or so, are continuous surfaces of polymer prepared using a molding process that in some chips employs the channels in the chips as molds. Using a channel together with a polymer permits varying the surface areas that contact the flow. If the polymer fills a large portion of a channel's cross-section, more active surface encounters a smaller moving stream. Increasing the probability of contact increases the likelihood of more molecules leaving highly dispersed samples thus facilitating more complete extractions. Additionally, we may alter the porosities of the polymer if we employ differing combinations of porogenic solvents and change the conditions for the reactions (Ref: Svec and Fréchet).

Changing Patterns of Flow

Circulatory patterns within a hemocoel are complex. Pathways for flow vary for many reasons including volume of hemolymph in the cavity, what enters from the alimentary canal, the bee's state of hydration, temperature, level of activity and developmental stage, to mention a few. For example, as muscles in the walls of the heart and dorsal vessel propel blood forward through the aorta, at the same time hemolymph leaves the dorsal sinus to enter the ostia of the heart. This forward flow in the vessel drags hemolymph in the dorsal sinus forward. At the same time, hemolymph moving backwards from the head courses through the visceral and ventral sinuses. Lateral and medial currents interconnect these sinuses and also pass around the alimentary canal as flows mingle over the insides of the external boundaries of the hemocoel. At the same time, some hemolymph, assisted by pulsatile organs and movements of the body, enters and leaves the wings and appendages.

Being Bitten

One can observe with considerable clarity how complex these moving flows can be if we are brave enough to watch an adult mosquito on our hand through a hand lens or dissecting microscope. As the mosquito's heart rapidly contracts, each pulse moves rapidly from the abdomen towards the head. At the same time, heaving movements of the internal organs, most of which do not appear to progress but remain at the segmental levels, mix the hemolymph flowing in the perivisceral cavity. As the gut distends with your blood or perhaps somebody else's if you are lucky enough, you may also see the ventral diaphragm pulsing in a posterior direction.

Packing Functions of Surfaces

In our devices, components surround a central silicon chip. Close configurations can provide power, remove heat and connect processors to other components. Packaging, consisting of three layers, supports the device. In the standard process for an Intel Pentium chip for example, droplets of solder pass current from the chip to the package. The grid of droplets or bumps connects to a network of copper wires in the top layer. These route to copper links that pass vertically through a plastic middle layer or core. On the bottom of the package core, the copper connections network through a third packaging layer to attach to larger pins that stick out from the packaging to connect to circuits on a mother board that link the processor to other components. This 'macro' design handles the forty-two million transistors on a Pentium 4 chip, but in biological packaging, elements having separate functions, such as nerves and capillaries, are much more closely associated at many different levels simultaneously.

Note on Reynolds Number

The Reynolds Number is a dimensionless combination of variables important in viscous flow studies for analyzing flows when there is a substantial velocity gradient or shear. The number indicates the relative significance of the viscous effect compared to the inertial effect and is proportional to the inertial force divided by the viscous force.

$$Re = \frac{D \cdot v\rho}{\mu} \tag{1}$$

or

$$Re = \frac{D \cdot G}{\mu} \tag{2}$$

where D = characteristic length, v = velocity, ρ = density, μ = dynamic (absolute) viscosity, G = mass velocity.

The viscosity above is dynamic viscosity also called absolute viscosity. For a pipe the characteristic length is the diameter of the pipe.

Chapter eight

SHRINKING

Overview of Shrinking

How do we shrink something? What happens to whatever's inside? May we shrink a larger machine down into its micro or nano version? Or must we organize small machines in entirely new ways? Shrinking alters things in unforeseen ways. Gravity affects small machines less for example. A flea jump dozens of times its height, but we cannot. How does smallness determine a system's operating speed, its power density, its output and overall efficiency?

Scaling laws are a simplistic indicator, of why nanotechnology may be extremely powerful even when compared with biology. Scaling laws let us compare the relative performances of systems having different scales to combine properties such as how a system's power relates to the system's volume giving us its 'power density.'

Consider a muscle inside a bee's leg. As we shrink the leg and muscles, strength decreases as the cross-sectional area of the leg and muscles decreases. A muscle's weight is proportional to its volume. Strength versus weight crudely indicates a muscle's power: length squared divided by length cubed or length to the minus one. Therefore, strength per unit weight improves ten times as our bee shrinks ten times smaller.

Consider gravity: A nanomachine, nearly a million times smaller than a flea, is entirely unaffected by gravity. We cannot compare directly strength and mass, but together both determine our shrunken system's performance.

Decreasing Size Increases Power

How about speed of response? A bee moves its wings up and down far faster than we can flap our arms. The speed of a moving appendage may be about the same in the bee and us, but the bee's wing travels a much shorter distance. So in small systems the speed of walking and flying movements increases. Think of a mosquito's whine. Or a factory might perform ten steps a second, but fast enzymes operate about a million times each second. The density of power or power density measures power, force or strength, times speed, but strength is proportional to area. If speed is constant, a machine ten times larger can produce one hundred times as much power. But power per unit volume or power density remains unidimensional!

Think of it this way. If a system ten centimeters cubed creates a thousand watts of power, then an engine one centimeter cubed may produce only ten watts of power or one hundredth the power of the ten times larger engine. But are you ready for this? Suppose a thousand of the one centimeter cubed engines occupies the same volume as the one ten centimeter cubed engine, so now together, the smaller engines create ten thousand watts. So by building a thousand times as many machines and by making each machine ten times smaller, the same mass and volume can be designed to deliver ten times as much power.

Suppose we now consider frequency of operation. Frequency of operation increases as the size of our system decreases, so miniature engines may run at ten times the rate of smaller ones. When a design shrinks by a factor of ten, its number of parts increases by a factor of one thousand. This relationship is the functional density, and functional density remains proportional to our system's volume. We can pack in a million, million, million, or 10^{18} more nanoscale parts that are a million times smaller into the same volume.

Suppose we pack these parts into a bee's thorax. Shrinking by a factor of one hundred, as might be the difference between today's transistors and today's molecular electronics, allows us to confine a million times more circuitry within the same volume. But here's the rub. Suppose each additional component costs extra money, or our parts or machines have short lives, then taking a thousand times more parts to just increase our current performance ten times becomes non-economical.

But what if we might coerce bees into producing our parts using their evolutionarily honed, massively parallel, reliable and fault tolerant, processing designs, so that our parts, now made by contented bees, might last as long as a bee itself? It might now be worth attempting.

And lastly, let's consider efficiency. A large-scale system that is ninety percent efficient may grow well over ninety-nine point nine percent efficient if we shrink it into the nanoscale and reduce its speed to keep its power and functional density constant.

But what about friction? After all, isn't friction the bane of all machines, even nature's? That friction is proportional to force or area implies that frictional power must be proportional to what power is consumed, regardless of scale. Let's say the thickness of a protective cuticle that is available to erode through rubbing and attrition decreases as we shrink our bee system. What happens? Even when very thin, the covalent bonds of cuticle remain strong enough to resist forces between sliding surfaces that are smooth and clean, so frictional wear alone should not break these bonds, as rubbing will never generate enough heat or force to break covalent bonds. Remember I have said nothing about chemicals. Most systems will not shrink all the way down to the nanoscale, however, as problems having to do with how the system's parts connect, intervene.

Building Smaller Machines

How might we design and build smaller machines? First, we observe nature and abstract away from nature our ideas of what we want to build. Were we to copy a bee, we must "transliterate" from the biological bee blueprint language into a machine language of our own. Think of what a bee does. Whether the hemocoel contains much hemolymph or merely moistened surfaces, connectivity within the hemocoel suffices for particles to move from place to place without obstruction. Flows and transport differ in large and small hemocoels.

We begin with generalizations: our common abstract notions of size, weight and scaling. Objects seem lighter in proportion to their size as we scale them down. When descending much below bee size, however, physics changes markedly. Something thrown, flicked or kicked instead of landing close to its actuator now usually ends up far away. Compare our jumping flea's ability to jump more than twelve body heights with our Olympic records for the high jump.

Or look at gold. Nanoparticles, having widths of a few nanometers to a few hundred, contain tens to thousands of atoms and live and have their being within the realm of mesophysics. At intermediate sizes, nanoparticles straddle quantum and Newtonian realms: realms where common elements often display novel properties. Of two nanoparticles of gold, for example, the slightly larger particle may melt at a different temperature and possess a different conductivity and be a different color than the smaller one. So at the meso-level of organization, instead of changing the components and composition of our materials, we might consider altering size (Ref: Size Relationships).

Insect Scaling

Insects range in size. Bees live towards the middle of this range. The smallest insects are the Ptiliidae or feather-winged beetles. Less than a millimeter in length, these beetles can crawl through the eye of a needle. Smaller yet are the Mymaridae, a family of small parasitic wasps. A male *Dicopomorpha echmepterygis* lacks wings, is blind and is about one hundred and forty microns long. The largest insects now are fossils. A Phasmid, or walking stick, from the Carboniferous, 375 million years ago, was about nine inches long, and a dragonfly from the Permian, 285 million years ago, possessed a thirty-inch wingspan. In comparison, a modern day African Goliath beetle weighs around one hundred grams. The internal organs of even these diverse insects, however, are quite similar across scales.

Scaling is basic, but we often do not think enough about what scaling means. Our mathematical formulas adjust for scaling effects, but insects and micro-machines live bizarre lives. Peeling a charged surface off another charged surface is sticky but easy when the charged surface is a sweater off a lover, but electrostatic attraction holds our miniaturized parts during production of a MEMS so firmly together that peeling is not usually an option. And sticking's not the whole of it. Flimsy micro-parts puncture surface tension with great difficulty. Think of a water strider on a pond. Or consider rice grain-sized wheels that generate too little friction to bear much of a load. Small objects have large surface areas relative to their masses, so the smallest insects having the thinnest cuticles may have difficulty retaining their internal moisture, and so they must live under constant threat of desiccation. A similar relationship exists for heat loss and heat gain. The list is long.

Scaling and Differential Shrinking

For example, how fast an organism takes oxygen from the air, digests and absorbs food, and loses or gains heat are proportional to the areas of the lung, gut, and body surfaces, respectively. These combined relationships mean that rates of acquisitions or losses are proportional to the masses or volumes of the body. They are also proportional to how the different functions connect. Or at any given mass, size is limited, because as skeletons decrease their masses accommodating other functions within a fixed volume, there comes a point when the skeleton becomes too flimsy and cuticles too thin to support the body. Skeletons are usually no more than ten percent of a system's mass.

Because the geometries underlying different functions change at different rates as an object shrinks, a change in the size of the object alone implies that functions related to size must also change, but at varying rates. If an organism or device is to remain functional as it shrinks, either these relationships must change, or the shape must change.

Size changes compel adjustments on many levels of organization simultaneously. Molecular forces, such as cohesion, become more important as mass decreases; flies walk upside down on ceilings because the force of gravity on a fly is less than cohesive forces holding the fly on the ceiling. Or to quote from Thompson's *On Growth and Form*: 'A man coming wet from his bath carries a few ounces of water and is perhaps 1% heavier than before; but a wet fly weighs twice as much as a dry one and becomes a helpless thing.' (Ref: Thompson, 1942).

Lilliputian Physiology

To appreciate the effects of scaling and shrinking on an organism we understand better than insects, let's now perform the thought experiment of shrinking a woman until she is about an inch tall. Remember, her body circulation won't circulate at these dimensions, but let's imagine we got around this problem and we can keep her alive. At this point her linear dimensions have shrunk by a factor of about seventy. Thus, the surface area of her body (through which she loses heat) has decreased by a factor of 70×70 or about 5000, but her body's mass (that produces heat) has decreased by $70 \times 70 \times 70$ or 350,000 times. As a Lilliputian she now has great difficulty maintaining her temperature. When her environment cools in winter she dies, unless her metabolic rate or heat producing capacity increases drastically.

The relative importance of physical forces working on and in her body depends on her size because of scaling again. How does she breathe? The surface area of her lung has only decreased by five thousand-fold, so she can still acquire the increased oxygen she needs perhaps by breathing more rapidly, but taking more breaths each minute challenges her musculature and requires more energy each minute to contract her muscles. Being so small, now like a shrew, she must eat her own weight in food each day just to stay alive and even more calories to support her increased activity.

Because our Lilliputan's surface area is now relatively larger after shrinking (she now has an increased surface to volume ratio), she looses water at a faster rate, so she must drink more. But now water's surface tension has become a major force in her life. When she was larger, surface tension was less than gravity. But drinking is now very difficult. The surface film tends to pull her into itself when she tries to drink. As a small 'machine' now living in the realm of microphysics, it would behoove her to develop long mechanically advantageous jointed stilt-like legs as well as a straw-like proboscis she could unroll like a mosquito's to poke a hole in the water surface. Her muscles must now attach differently than when she was bigger.

Consider the problems of a recluse spider. The spider's jaws clamp as it bites with its chelicerae at a force that is proportional to the cross-sectional areas of its jaw muscles. However, the spider's weight is proportional to its volume. So to bite and puncture human skin is difficult for the spider. Our Lilliputian would be at similar mechanical disadvantages, but she has also gained an advantage. Being so small, like a little spider now, she falls gracefully through the air.

A falling Lilliputian accelerates until the drag force imposed by the air on her body equals the gravity acting on her mass. When these forces are equal to each other and from this point on, her falling velocity is constant. Her now constant falling speed is her terminal velocity, and for an adult person, terminal velocity is close to one hundred and twenty miles per hour. Air's drag on a moving object is proportional to the object's cross-sectional area, but the force of gravity is proportional to the object's mass (and thus volume, if the density is constant). As objects shrink, gravity's pull decreases more rapidly than drag, so terminal velocities of small objects decrease.

A falling object acquires kinetic energy proportional to its velocity of falling squared. Kinetic energy dissipates rapidly when the object hits something and stops falling. A falling spider's health is potentially better than a falling person's. Smaller objects fall more slowly, but because of the squared velocity term in the kinetic energy relationship, much less energy dissipates when these objects impact, so injuries are less. An elephant falling out a window will be hurt or die, a squirrel may not be hurt, but our Lilliputian runs away unharmed, that is if her heart is up to it.

Connectivity Maintained

To shrink a system and keep it working in its shrunken version requires that we maintain our system's connectivity as we shrink it. Ideally, we must follow what happens to the networks of communication within the system. For example, endocrine glands secrete hormones at one location, and these chemical messengers circulate to other parts of the system where they attach to receptors, eliciting effects at a distance from their sources. What happens moment-to-moment depends upon the rates of hormonal production and utilization as well as how well the circulation continues to maintain balance and distribution so that the shrinking system remains effective. Production of hormones might have to slow as shrinkage occurs, for example, to compensate for decreasing masses of controllers to remain in balance with their controlled tissues. Without having to understand the molecular details of any given network, we can grasp a network's common architecture or its scale-free topology.

Scale-Free Topology

A scale-free topology means that the probability, $P(k)$, that an arbitrary element of the network connects to exactly k other networks has the form: $P = Ck^{-\gamma}$, where gamma is usually the scale-free exponent. Only a few elements of any scale-free network link with many of the other elements of the system, and most of the elements link with just a few elements. Such scale-free topologies undergo phase transitions from their ordered dynamics to chaotic dynamics as size change. Gamma and a system parameter, *P*, constrain scale-free topologies. The dynamics of these networks are largely unexplored (Ref: Scale-Free Topology and Robust Networks).

Implications of Scale-Free Structure

Understanding a network's scale-free structure has led to interesting results in diverse areas. For example, people believed initially that the best way to curb spread of a computer virus was to provide all machines with antiviral software to make them resist infection, but studies of random graphs of the Internet indicate that antiviral software in an increasing numbers of machines had a cumulative effect that does not occur in the scale-free setting of the internet. In scale-free settings, adding antiviral software to machines at a relatively small number of hubs of the scale-free system can stop spread of the virus completely.

This insight is analogous to the impression that a network of human sexual partners appears to be scale-free as well. We may slow or stop the spread of AIDS by treating people at the highly connected hubs, in other words, people having the most sexual contacts. So by considering a shrinking system such as a hemocoel to be a system of hubs simplifies thinking about how to shrink the system and how the system communicates with itself.

Hemocoel Dynamics

Consider a wave of particles moving through a compartment of a hemocoel as the hemocoel shrinks. Hemolymph continues to move and evolve through the intricacies of the architecture over the surfaces of the hemocoel as long as particles can find a connected path of permissible pathways that are above threshold all along their course. At any moment depending on position of the insect as well as yaw and pitch and vibrations of walking or flight, the fluid particles are either driven at random or start out from a random state, but for flow of information to occur between two points, regions able to transmit flow must remain joined together in a random network even though internal and external dynamics continue to modify the random network with time.

Percolation: Modeling the Randomly Connected Hemocoel

Imagine a hemocoel to be a porous brick wetted by hemolymph. We may ask what is the probability that the interior of the hemocoel is wetted so that a particle can move from one place to another through the interior of our hemocoel brick. Using ideas from Broadbent and Hammersley's percolation model (Ref: Percolation), we can move between three dimensions when the brick or hemocoel is fully saturated towards a two-dimensional model when the hemocoel consists only of moistened surfaces that connect with each other two-dimensionally.

In two dimensions, we have the following: Let the surfaces of the hemocoel together be represented by a plane square lattice. Let probability, p, be a number between zero, where the edge of a square is dry, and one, where the edge is wet. Now we examine each edge of the lattice in sequence and declare the edge to be wet with a probability p and dry otherwise. Each edge is wet or dry independent of all other edges. The edges of the two-dimensional surface represent the inner connectivity of the hemocoel so that p becomes the proportion of passages that are wet and thus allows flow to cross them. Grossly imagined, when $p = 0.25$, the connected clusters of open edges are isolated and small, but as p increases, the size of each cluster increases, and the number of clusters increases, until at a critical value of p, called Pc, a cluster forms that now fills the entire space of the hemocoel.

Now, we may model any sized hemocoel using a large finite subsection of this two-dimensional surface. As dynamics change and flows vary, for the hemocoel to transmit hemolymph and continue to function as a hemocoel, the vertices and edges of the square lattice must contain somewhere a connected subgraph of the surface. A point within the model is wetted if and only if a path exists in two dimensions connecting a wet square of the surface to our point in question. The hemocoel functions as a circulation as long as 'a wet line of squares' exists from point to point across the hemocoel, but connectivity does not depend on the length or direction of this line that may change moment to moment. Percolation theory investigates the structure of this subgraph when we delete the closed edges particularly in regards to how the percolation structure depends on the numerical value of p. If p is large the probability of flow is increased over times when p is small.

For purposes of shrinking, it is evident that the fine structure of the 'passages' in the interior of the hemocoel is on a scale that is negligible when compared with the overall size of the hemocoel. In such situations a vertex in the center of the hemocoel is wetted with hemolymph and, hence, can give up and receive particles from the circulation, behaves rather similarly to the probability that this vertex forms the end vertex of an infinite path of open edges on the wet surface. It is for this reason also that a hemocoel is robust and resists point blockages, as given a minimum of hemolymph to wet the surfaces, there are many alternative routes for particles to follow in case one path becomes obstructed. We may construct an analogous model for a three-dimensional volume. Our above model is called bond percolation on the square lattice and is the most studied of all percolation processes.

Transliteration: From Bee to Device

The bee system is uniquely compacted. Imagine a large bee circuit board before we shrink her. Bees are micro-packages; each bee's systems coordinate her into an assembly robot. Hordes of bee-robots in colonies assemble wax cells into combs for nurseries and storing honey, defend the colony and locate food sources. Each bee "package" resembles a very complex mass-produced chip, comprising a system of sensors, a coordination center and effectors. Parts on her chip are her compound eyes, a brain like a microprocessor and the cantilevered actuators of her wings and legs. She senses her environment. Her CPU or central processing unit compares inputs, chooses outputs, actuates actuators, and like any well-behaved robot, she responds to standard stimuli in quite predictable ways. Using no tools and at ambient temperatures she recreates herself. Bee-manufacturing has miniaturized her fuel supplies, sensors, CPU and actuators, while employing no toxic metals or high energies but only "clean" biological sources of energy. Thus, bees shrink and integrate far better and more economically than we can.

Although we have studied the structural properties of many such networks including a bee's anatomy quite thoroughly, most dynamical properties of these same networks remain unexplored.

Geometrical or size correlations of function depend on each system's geometry, architecture and materials. Interactions and relationships between size, structure and function across different scales are unknown. For example on 23 February 2005, searching for the word 'insect' together with the word "mesophysics" in Google did not match any documents.

Trend Towards Miniaturization

Our manufactured systems have shrunk over time: radios, pumps, typewriters, motors, so that today an electric motor sixty microns or sixty millionths of a meter across or less than a hair's breadth wide, can still spin when properly electrified. But static electricity poses difficulties when manufacturing anything this small. To shrink something in the lab is difficult but not impossible, but manufacturing our something in large enough batches is the rub. Our technology, great that it is, still limits our building small. We can either fit small parts together to make larger pieces, or we can make flat things all of one piece on chips.

Remember How to Build a Chip?

Chip manufacturing illustrates some of the problems we encounter while building small. CMOS, pronounced sea moss, stands for complementary metal oxide semiconductors. Conceptually, making a chip is like designing and printing an etching we have designed on silicon. First conceive of and draw a pattern for the connectivity of the components. Then project this pattern as lines onto a wafer of crystallized silicon. Then with a caustic chemical etch away all the places we do not want silicon. Then deposit one layer of metal and semiconductor and then repeat the etching deposition process layer after layer until we build up an elaborate three-dimensional structure of trenches, columns and basins, each filled with metals and semiconductors. Each layer, however, retains its unique two-dimensional pattern confined within its own layer. To create three-dimensional mechanical parts from our 'sandwich,' we etch material away around our pieces until parts are thick enough to withstand the stresses and strains of working together as a MEMS. We can fashion from our cutouts, for example, diaphragms for micro-pumps and blood pressure sensors that flex when under pressure as well as cantilevers that work as accelerometers to activate airbags. But undoubtedly over time, all our 'pieces' will chafe because friction wears them away where they rub on each other.

Mesophysics and Granular Models

Now lets continue shrinking until our system is smaller yet. The realm of mesophysics currently incorporates microelectronics, quantum computing and molecular biology. Mesophysics asks such questions as how do we transport electrons and how might we achieve higher densities of components in our microelectronic silicon-based circuits. In conventional circuits, electrons flow as would streams of water confined within the boundaries of their conductors. The wider a conductor is, the less the conductor confines a carrier and the more continuous the flow. If we shrink a conductor's diameter as much as possible, we must ask how do electrons squeeze past the narrowest places? Because of the coulomb blockade effect, electrons can distribute themselves to pass in single file, but a conductor's shape and properties and how electron motion and quantum fluctuation interact create many complexities we do not yet understand. But we have several analogies.

Analogy: Sandpile

Consider how a sand pile behaves. A sand pile is a simple system whose components interact by exchanging forces or information. Gravity drives the sand pile system externally. The grains represent flowing molecules. Piled sand grains display unique behaviors as we add grains to bring the pile into unstable, non-equilibrium 'avalanche' conditions. The changes grains undergo as they move in an avalanche models the solid to liquid transition because the avalanche recruits surrounding grains. Temperature plays no role in the dynamic transitions between phases, as the system responds with its own nonlinear dynamics to applied forces (Ref: Self-organized Criticality).

Self-organized criticality (SOC) embodies the idea that complex behaviors develop spontaneously in certain many-body systems whose dynamics change rapidly. Self-organized criticality in some basic way may influence the development of structure in biological systems. What are the properties of systems that lend themselves to similar sand-pile like cascades? These systems require events to occur on separated time scales for one. What constitutes the external driving of the system must occur slower than the time for the system to relax. An example would be the shaking of hemolymph from the movements of flying and walking and the imparting of the energy of body movements to moving the hemolymph (Chapter 9). Signals of any form traverse the system as long as a signal encounters a connected path of above threshold regions for its propagation. Making analogous granular models even of some of a hemocoel's dynamics, might suggest how information might circulate. Investigating the consequences of the inherently inhomogeneous distributions of forces inside piles of granules might allow particles of a system to explore phase space as well as other aspects of aggregation. Looking at the phenomena of aggregation and the spontaneous formation of structure together with changing instabilities under applied stresses suggests how energy and momentum might propagate through highly dissipative materials. One might use video, magnetic resonance imaging (MRI), and X-ray tomography to record a model system.

Analogy: Microfluidics

The sandpile analogy suggests that in the smallest systems microfluidics, the study of microflows, must be central to any ideas we might have for shrinking complex insect-like systems (Ref: Microfluidics and Drops). A dripping tap reveals just how complex even the simplest microfluidic system is. Singularities in patterns of flow occur almost everywhere, even in free-surface flows.

A drop hanging and then falling from a faucet models a liquid separating into two or more pieces. The change in structure of the drop over time reveals that as the drop falls, it first tugs out a long neck between two masses of fluid. One mass will remain on the tap; the other will fall as a new drop. The neck thins out elongating until it breaks. What is the shape of the drop as the neck fractures? Something complex happens in the mathematical description of the liquid at this critical point, because the drop undergoes a transition in its topology. A drop starts as a single, connected fluid but ends as two or more separate drops. Separation is just one example of a finite-time singularity, because a drop's breakup happens just after the drop becomes unstable enough to fall. A singularity arises for the topological transition because the radius of the neck holding the drop to the larger mass of fluid gradually shrinks and disappears. As the radius of the neck goes towards zero, the drop's curvature diverges, so the forces of surface tension become infinite. How do such dramatic dynamics develop in a system that has smooth initial conditions and forcing terms? Similar transitional singularities are common in many diverse systems, from stellar structures to turbulent air and water to a bacterial colony's growth. Now imagine what happens to hemolymph in the hemocoel of a bee on the wing!

Lab-on-Chip Devices

Understanding microfluidics facilitated developing integrated lab-on-a-chip (LoC) devices we use now for clinical diagnostics and to screen minute quantities of dissolved compounds. In most microfluidic devices, continuously flowing liquids traverse micro-channels fabricated mainly from glass or plastics. Both LoC devices and the bee's hemocoel rely on surface tension to manage discrete droplets at small length scales. One technique involves ElectroWetting-On-Dielectric (EWOD) where a varying electrical potential changes the wettability of liquids placed on the surface of a dielectric. EWOD reduces a sample's size below those required for conventional, continuous flow microfluidic chips, as well as reconfigures and rescales the chip's architecture. The similarity of the EWOD system to digital microelectronic systems has engendered 'digital microfluidics.' Now EWOD is used to dispense, cut, and transport tiny droplets.

Analogy: Crumpling

Crumpling also plays a role in shrinking surfaces and folding them into smaller volumes. How does a fold, or in simple terms, how does a sheet of paper crumple into a small ball? If one compresses a crumpled sheet of paper into a very hard ball by hand, nearly eighty percent of the ball's volume still remains air. What gives the crumpled ball its strength? Analogies are endless.

Argument

I argue that as a hemocoel shrinks, reduced volumes of hemolymph compared with the volume of the cavity are to be expected in the smallest insects, because as bodies shrink, a three-dimensional volume of hemolymph reduces towards a moistened two-dimensional layer on the external surfaces of the organs and the inner surfaces of the hemocoel. Reducing the volume of liquid hemolymph reduces overall weight and the energy expended to transport the weight, as well as increases the probability that molecules diffusing in the surface films of the hemocoel reach their destinations.

Empty Cavities

Smaller insects tend to have smaller volumes of hemolymph in the spaces of their hemocoels than do larger forms, but careful comparative studies of the relative volumes in different species are scanty. The volume of hemolymph varies with the species, age, diet, developmental stage, activity, and how volume is determined. Methods of collecting hemolymph in the past have included exsanguinations, dye dilutions and C^{14} measurements. Each method presents different technical difficulties and can be unreliable, but exsanguinations and dye dilution in trained hands can yield similar values (Jones, 1997: p. 74).

Ancillary evidence shows that some nymphs, termites and many caterpillars contain volumes of hemolymph that are large enough to create positive transmural pressures in the hemocoel; in some even enough hemolymph under sufficient pressure to swell their bodies. Many larvae maintain a constant volume of hemolymph relative to their weight throughout larval life, and these larvae continually produce hemolymph as they grow. Neurosecretion of hormones by the brain stimulates the corpora cardiaca to produce a diuretic hormone, but these endocrine interrelationships are complex. Removing the neurosecretory cells by excising the corpora cardiaca and allata may not affect water balance in some species. Fluid volume in an insect may peak prior to molting, but then this volume drops quickly afterwards. Males may possess less hemolymph than females. But in general, it appears that the volume of hemolymph may be smaller in proportion to the size of the cavity in smaller insects (personal observation).

In many insects, especially the smaller ones, cavities appear 'empty' and only 'wet' inside when opened. One exception to our size rule are the seventeen-year cicadas captured in 2004 in Maryland. These cicadas, had large moist empty hemocoels and very little fluid hemolymph. As in these cicadas, microscopic preparations of the smallest forms, when carefully fixed and sectioned for light and electron microscopy, present little evidence for accumulated fluid.

Shrinking Electronics is Easier Than Shrinking MEMS

Problems of shrinking devices arise from many interacting factors. Gordon Moore in the 1960s noted that the number of transistors on a chip doubled every one and a half years. This exponential growth is now Moore's Law (Moore, 1979). Moore's Law is not an immutable law of nature but one describing the results of summed feats of exceptional engineering that make it possible, perhaps, to predict the limits constraining the future sizes of our very large-scale integrated circuits.

The Bottom Line

To be successfully shrunk, either as a bee or a device, and to still go on working, increase your surface-to-volume ratio. Internal organs or parts of devices extend surface areas relative to a system's volume. Lungs increase surface area for the exchange of gases; the circulatory system distributes material to an internal space that cannot be reached by direct diffusion from the external surface of large organisms; intestinal vili increase the surface area available for intestinal absorption. A tapeworm that lacks a circulation, may be twenty feet long, but a worm may not be thicker than a few millimeters, because food and oxygen have to diffuse through the skin to reach all the cells of its body.

Gravity's relative weakness at insect sizes permits insects to have thin, external skeletons permeable to many molecules. The disadvantage of wearing a corseting cuticle is that it enables growth only when sloughed and a new one forms to accommodate an enlarging body. Between shedding and regrowth of cuticles, the insect body remains soft. Were mammals to do away with their skeletons, even for a short time to change skins, the organs of mammalian bodies would collapse under gravity. Lobsters and crabs can grow larger than insects do in air, because water buoys them up so these arthropods spend their 'soft' stages nearly weightless.

Large terrestrial organisms, all in all, resemble each other by having thick legs and relatively short, stout bodies. The "invention" of building internal organs so that they increase their surface areas, helped animals over time to retain their highly successful simple exterior shapes to house large internal volumes. The principle of maximizing surface to volume ratios is important for the nesting of functions on surfaces within devices. Limits may be expanded, but these laws still operate. No Gothic church may be higher than it is long, and no large animal may dip in its middle like a dachshund.

Chapter nine

CHANCY TRANSPORT

Overview

Cooling small computers or miniature systems requires novel cooling systems. Many could employ Brownian motion with coupled convection as in heomocoels. Can we learn from bees and insects how they cool themselves, so we then may incorporate insect techniques into our devices?

We now examine how things move around through hemolymph and the hemocoel. Topology, geometry, and microphysics influence how particles distribute themselves. Diffusion, a metabolically cheap transport mechanism, is the process where random molecular motions move matter from one part of a system to another. However, diffusion's effectiveness decreases rapidly with distance. An advantage of small hemocoels, then, over larger ones might be that they increase diffusion's effectiveness. Pumping of the heart and movements of legs and wings supply convection and perhaps even superdiffusion.

Were we with a light microscope to watch visible particles small enough to share the molecular motions of the hemolymph, we would see them move randomly. In a dilute solution, each particle collides with molecules of solvent but behaves independently of the other solute particles that it seldom encounters. As a result of so many collisions on all sides, particles have no preferred directions, moving sometimes towards a region of their higher and sometimes towards a region of their lower concentrations.

Our conventional, familiar assumption is that concentration differences determine rates of diffusion and that steep gradients create faster displacements. However, this simplistic assumption only approximates a much more complicated situation (Ref: Diffusion).

Our particles proceed by random walk or Brownian motion, so that in a quiet hemocoel we may calculate the mean-squared distances each travels in an interval of time, but we cannot foresee in what direction any particle will travel. Brownian motion, due to bombardment of particles by the thermally excited particles of their solvent, was one of the first natural geometries recognized as being a self-similar fractal. If we view a finite segment of a two-dimensional random walk and then re-scale its time and length, we see that the two patterns resemble each other and are self-similar.

Now imagine particles in the hemolymph of a bee following cardiac arrest so the hemolymph is still and two separated locations in the hemocoel. Particles might be nutrients entering the hemolymph through the wall of the midgut that are diffusing towards a cell in a flight muscle in the thorax that will absorb and then consume them. In addition, let us assume a gradient of concentration for our nutrient molecules, such that the particles exist in higher concentration near the wall of the midgut, so they diffuse away from the midgut towards sinks in the thorax muscles. These particles on the average progress from a region of their higher concentration to regions where they are less concentrated.

Concentration Gradient

To understand how net movement of particles occurs, consider an imaginary horizontal section taken perpendicularly across the concentration gradient, and imagine two thin, equal elements of volume, one downstream and one upstream, from our horizontal section. Though we cannot say which direction any particle will move in an interval of time, we can say that on the average, a definite fraction of the molecules in the upstream element will cross the section moving down and that the same fraction of molecules will cross the section moving up in the same interval. Because the concentration of particles is greater in the upstream element, however, a net transfer of nutrient particles occurs from the uphill side where nutrient particles are in greater concentration towards the downhill side where nutrient particles are in lesser concentration. If nutrients continually enter the hemolymph from the gut and are taken up by muscles in the thorax, our driving gradients will transfer mass and energy directionally through the hemocoel even though all particles move by random molecular motions.

Fick's Law and Diffusion Coefficients

Fick's law states that the flux of nutrient particles across our imaginary planar section due to their random motion is approximately proportional to the local gradient in the concentration of particles. Together with a term, dc/dx, that describes the change in concentration over distance, the diffusion equation depends on D, a diffusion coefficient, whose magnitude does not depend on distances or concentration but instead describes the mobility of particles as a function of their size, charge, the nature of their interactions with the solvent, and the temperature.

Without solving the diffusion equations, we may say that the units for the diffusion constant, D, are distance squared per unit time. This observation alone says that the average distance through which diffusion operates in an interval of time is proportional to the square root of the product: D times time, and that the time taken to diffuse a distance, d, is proportional to the distance squared divided by the diffusion coefficient.

Diffusion Coefficient: Example

A typical small molecule is glucose. Glucose in watery hemolymph might diffuse ten to the minus five seconds centimeters squared per second. Consequently, the time glucose needs to diffuse one micron is ten to the minus three seconds, to diffuse ten microns is a tenth of a second, and to diffuse one millimeter is ten to the third seconds or fifteen minutes. Diffusion is rapid at intracellular scales, so cells need not add energy to this metabolically free transport mechanism, but diffusion alone is inadequate over distances greater than one millimeter. So remember; if you are larger, you need a heart.

Geometry Influences Diffusion

Transit time depends heavily upon dimensionality. Flat, thin animals can live by absorption and diffusion without convection. Tapeworms do not require a circulatory system and can elongate without altering diffusion distances. In one dimension, transport by diffusion is proportional to distance. At three dimensions, however, the multiplicative factor is distance divided by the radius of the sink consuming the particles (Ref: Hardt, 1980). Therefore, to speed transit in larger forms, we must add convection, and this means adding a circulation with a heart or two or three.

Adding Convection

Now imagine a bee's heart pumping hemolymph as she subjects her body to the pitching, yawing and impacts of walking and flying. Consider also that her hemocoel contains only a small volume of hemolymph, a volume just sufficient to coat the surfaces of her internal organs, the walls of the hemocoel and her diaphragms together with perhaps a small pool of hemolymph in a corner somewhere. Now her hemocoel is a connected two-dimensional surface within which our randomly diffusing particles move.

Somewhere in shrinking a larger volume of hemolymph to a smaller volume, a three-dimensional volume approaches a two-dimensional volume so that transport now depends more and more upon the moistened geometries of a connected, almost minimal, two-dimensional surface.

The Two-Dimensional Hemocoel

We characterize our two-dimensional hemocoel in different ways. If it contains many randomly moving molecules engaged in Brownian motion, we might imagine our 'random walkers' to traverse fractal landscapes. We understand something of fractal elastic properties as well as phase transitions occurring on fractal structures, but we know very little of why fractals form in the first place. We don't even know which aspects of the evolution of a dynamical system give rise to fractals (Ref: Fractal Difficulties).

As hemolymph disperses over irregular surfaces, at the micro-scale we might see shallow pools around higher, dryer islands. Or our simple picture of diffusion as a sequence of random steps changes if our hemolymph resides in and percolates through a porous medium having a 'porosity' less than one. Under these conditions, our random walkers, now obstructed by 'islands' or pores, cannot occupy all positions of space. If the porous matrix is homogenous and isotropic, an effective diffusion coefficient involves a formation factor that now is no longer purely geometric, like the porosity, but has become a transport coefficient (Ref: Percolation, Chapter 8).

On free surfaces in some systems, gradients in surface tension drive convection depending on the Marangoni number, an equation relating the fluid density, its kinematic viscosity, the interfacial energy per unit area of surface or the surface tension and a characteristic measure of the size of the container, which in our case might be a hemocoel (Ref: Meakin, 1998).

We shall continue this discussion after again considering the three-dimensional hemocoel of Deuterostomes.

Deuterostome Diversion

We now make the case for the large, three-dimensional hemocoels of sea urchins and starfish. In a three-dimensional volume, there is only about a thirty-four percent chance for molecules to traverse the cavity from one location to another without getting lost; so how can deuterostomes receive adequate energy if two-thirds of their nutrient molecules get lost in the infinities of their internal oceans of coelomic fluid? The answer must involve the idea that echinoderms, being exclusively marine, lead lives that are slow and cold enough for diffusion coupled with some minimal sloshing of fluid to supply their metabolism. Echinoderms move slowly and require less energy than faster moving forms. Beating cilia on the walls of coeloms may contribute a weak current.

For zoological reasons, the cavity of a starfish or sea urchin is a coelom and not a hemocoel, but coeloms work like hemocoels. However, all is not so simple in echinoderm larvae that possess gel-filled cavities (Ref: Gel-Filled Cavities).

To observe the movements and fates of particles in the coelom of a starfish such as *Asterias*, inject less than a milliliter of a carmine suspension in seawater into one of the rays of a living starfish through a hypodermic syringe attached to a twenty-gauge needle. See the red color rapidly traverse the ray to enter the body. You will see how the cilia on the walls of the coelom create an internal circulation.

After ten minutes, withdraw a few milliliters of coelomic fluid from the starfish and examine this fluid under a compound microscope. You will see that amoeboid phagocytic cells in the fluid are engulfing the particles of carmine. Along the margins of the radial canal (remember this is the hydrostatic system having to do with activating the tube feet) are nine areas, called Tiedemann's Bodies. These structures probably filter the coelomic fluid because they remove the carmine-filled cells.

General Assumptions

We reason that as cavities shrink transport grows more efficient because distances shorten. There are advantages to shrinking one's hemocoel. Fluids are heavy, and a shrinking animal loses weight, but diffusion through watery films insures transit. Confined to a plane surface and given infinite time, a randomly moving particle covers the entire area, so the probability of its transmission is now one, and delivery is assured provided we wait. Movements may also agitate the hemolymph, but the countering forces of cohesion and adhesion can offset effects of jostling, but in some areas superdiffusion may operate. The hemocoel, within limits, adapts to changes of volume, and in small bodies, supporting struts of a reduced skeleton may in turn be less massive, because fluid films behaving as minimal surfaces reduce total surface tension traction on skeletal supports (Ref: Lungs and Struts).

Diffusion Within a Plane Surface

On a plane, all diffusing substances enter the film of hemolymph through the faces of the film, and very little enters through the edges. If the thickness of the sheet and the diffusion constants do not change, a steady state is reached in which the concentration of our solute molecules becomes uniform throughout the sheet, so that any difference in concentration with distance is zero. Now again, imagine two locations separated by a distance with a gradient of concentration for our substance existing between them. Our once three-dimensional volume is now sandwiched into a film. As before, the difference in concentration determines how fast transfer occurs between our two points. (Simple experimental arrangements for measuring diffusion coefficients in planar situations are in Newns, A. C. (1950). *J. Tex. Inst.* 41: T269.)

If we imagine our sheet of hemolymph to be a thicker sandwich of superimposed layered films, then the fall in concentration through the sheet is the sum of the falls through each layer, and the resistance to diffusion through the sheet is the sum of the resistances of its separate layers. We must assume, of course, we have no diffusion barriers between the layers.

Plane Surface Contacting a Stirred Pool

If a sheet wet with hemolymph contacts a large, well-stirred volume of hemolymph, the amount of solute that leaves the pool for the plane surface is a negligible portion of the total solute available, and the concentrations in the pool and sheet may remain constant as diffusion proceeds. (Equations and references for two-dimensional solutions for these and similar situations are in Crank, 1975.)

Role of Mechanical Dispersion

Fluctuations in local velocity of flow disperse particles mechanically. If at the macroscopic scale, the fluctuations do not correlate, each mechanically perturbed diffusing particle follows a tortuous streamline of the flow we assume would otherwise be straight. During convection, in any interval of time, particles distribute more widely and deeply in both the horizontal and vertical directions. Because only convection creates mechanical dispersion, convection can occur when diffusion is absent.

Boundary layer dispersion interferes with diffusion when particles move slowly near solid boundaries. In experiments we can observe tracer molecules to enter and leave boundary layers both by diffusion and convection. When fluid moves slowly, diffusion is the only mechanism moving particles into or out of boundary regions. Hydrodynamic dispersion may occur as flows stagnate. Taylor dispersion results as convection combines with molecular diffusion (Ref: Mechanical Dispersion).

Circulation Time

Hemolymph does not always circulate in our usual pattern. Witness the mosquito feeding on your hand. Observe the pumping abdomen filling with blood and the heart rapidly contracting forcing blood forward as moving abdominal viscera and a distending blood-filled midgut stir the hemocoel. Under a microscope you can see ripples in the ventral diaphragm progressing backwards.

In embryonic and pupal stages, even after hearts develop, hemolymph may not circulate rhythmically, but hemolymph circulates before the dorsal vessel forms. Once the dorsal vessel begins to pump, however, flow follows the generalized dorsal, lateral and ventral pathways previously established around organs and skeletal elements. Attachments of the massive muscles for walking and flight complicate the patterns of flow through the thorax. The usual pattern, providing impetus for flow, is that the accessory pumping organs, contractions of the alimentary canal, respiratory movements, and contracting muscles and other organs assist the heart.

Cardiac Output

What is a bee's cardiac output? How long it takes for hemolymph to make a complete circuit depends on how fast the dorsal vessel beats, the volumes of hemolymph pumped in an interval as well as other factors. Multiplying the volume per beat and the beat frequency gives an estimate of the volume of hemolymph that traverses the dorsal vessel in an interval or the cardiac output. Activity, whether walking, flying or resting, the developmental stage, hydration and temperature, play contributory roles. Few measurements of circulation time through a hemocoel exist. Esch in 1960 observed a circulation time of two minutes in an adult bee (Ref: Esch, 1960).

Lattice-Boltzmann Model

Let's ignore for a moment any complications of locomotion. Flows over surfaces within the hemocoel are complex enough.

A Lattice-Boltzmann simulation may reveal interactions on the microscale, and a simulation would allow modeling the microfluidic dynamics. Wettability of the surfaces, phase interfaces, and chemical properties interact to determine transport of momentum, heat and mass through the hemolymph.

Using a Lattice-Boltzmann model, we divide the hemocoel into a regular lattice and assign a set of velocity vectors to each lattice point. To connect each lattice point to its neighbors, we assign specified magnitudes and directions to our vectors. To define the total velocity and density of fluid, we specify how much fluid moves with each vector in each interval. By using time increments, we evolve a fluid distribution function that moves our particles progressively stepwise through the hemocoel. We simulate how particles collide by relaxing our distribution towards an equilibrium distribution having a linear relaxation parameter. We must specify rules for these interactions, so they satisfy laws for the conservation of mass and momentum to give a second order solution of the Navier-Stokes equations. Similarly constructed Lattice-Boltzmann models simulate heat transfer and phase changes in solid-liquid interfaces along micro-channels of micro-fluidic devices. Some include the effects of wetability on wall slip (Ref: Lattice-Bolzmann Models).

Convection

When the temperature of a bee or device at heat equilibrium remains constant, heat production equals heat loss. Input of heat from solar radiation or metabolism or resistances within devices equals the 'leakiness' to heat (heat conductance) times the difference between system temperature and air temperature. In the bee, this conductance measures how much oxygen is consumed to maintain heat balance, expressed as oxygen demand or milliliters of oxygen consumed per weight of bee per unit time.

Small devices or insects may maintain lower temperatures than larger ones. Small systems often produce heat at higher rates, but the higher rates require higher conductances, or put another way, smaller systems require increased circulation of coolant because of their larger relative surface areas. Large bees heat up while flying and elevate their body temperatures, even though the metabolic cost of flight per unit weight drops about two hundred and thirty percent for a tenfold increase in mass.

To increase the heat capacity of a system and decrease variations of temperature it experiences, its coolant must absorb large quantities of heat over a narrow range of temperatures. If a bee or device is small, and its surface to volume ratio is large, it either changes temperature with the surroundings, losing or gaining heat by convection as needed, or the system burns energy to cool or warm itself independently of the environment. In biological systems, enzymes catalyze metabolic processes that are efficient only over small ranges of temperature.

Cooling small computers or miniature systems requires novel cooling systems. Can we learn from bees and insects how they cool themselves, so we then may incorporate insect techniques into our devices?

Consider any small system. Because hemocoels dissipate so much energy from active flight muscles, hemocoels must remain cool. Because distances between interior points in the hemocoel and exterior points on the body's surface are so short, conductive cooling often suffices. On the other hand, if a hemocoel is to operate when a bee is cold, she may have to linger in the sun before taking off.

Warming Hemocoels

Watch bees at a hive entrance on a cold morning. Lethargic bees beat their wings slowly and stiffly. Insects balance the heat they absorb and produce against the large quantities they must lose to the environment, but at early morning start-up, a little sun helps. Insects control heat gain or loss or thermoregulate depending on need. Contracting muscles release heat. About ninety-four percent of a muscle's energy degrades into heat; only the remaining six percent produces mechanical force.

Hemocoels distribute heat by conductive cooling or warming. Suppose for a moment the hemocoel were a connected tubular system supplied by a pump. For any given rate of flow, tubes of large diameter would present less resistance and strain to the pump, but a large tube linking a heat source to a heat sink would be inefficient, as it would make poor thermal contact with the muscular generators of heat. Giving the tubes smaller diameters might improve thermal contact and heat exchange, but the increased resistances of all the small tubes together would strain the pump.

So the pump tubular system of vertebrates is a compromise, employing tubes of large bore to carry blood over the greater distances but tubes of smaller bore to cover the remainder. Linking large arteries and veins to capillary beds satisfies Murry's Law. Murry's law states that for a system of tubes containing laminar flows, the minimum volume for any given drop in pressure occurs when the sum of the cubed radii of the smaller tubes at a branch point of a vessel equals the cubed radius of the larger tube leading to or from the branch. This balance equalizes shear stresses in the tubes. For many fractal systems, tube lengths are approximately proportional to their radii, so the sum of the areas of the tubes at each level of the hierarchy conveys approximately the same volumes. Circulating blood resides for almost equal times at each level (Folkov and Neil, 1971).

Heating and Cooling Review

Consider a flying bee. At any moment her temperature is a balance between heat produced or gained from the environment and heat lost to the environment. The body's conductance or leakiness to heat is the difference between air temperature and our bee's body temperature. Body size, flight speed, wind speed, insulation and circulation of hemolymph help to determine her heat conductance.

Now what temperatures can she tolerate? If too cold, her thoracic flight engines scarcely 'turn over,' but if too warm, her enzymes denature and fail. Small bodies have lower temperatures during flight than do larger ones, even though small bodies may produce heat faster than larger bodies. The large surface-to-volume ratios of smaller bodies enhance convective heat loss better than do the smaller ratios of larger bodies. For example, a flying mosquito maintains a less than one degree centigrade gradient of temperature between its thoracic motors and ambient, whereas a flying bee heats its thorax to fifteen degrees centigrade above ambient. Her larger thorax means she has a smaller relative surface-to-volume ratio, so her convective losses are inadequate until her motors create a much larger temperature gradient.

Counter-current Heat Exchanger

Honeybees fly at ambient temperatures up to forty-five degrees Celsius when other 'bee-sized' insects cannot. Bees employ an efficient counter-current heat exchanger that apposes two oppositely directed flows of hemolymph in close proximity to control heat loss and heat gain to maintain heat balance. If one flow is of a higher temperature, heat passively flows downhill from higher to lower temperature across the wall separating the flows.

At warm-up, hemolymph flowing aft conveying heat from thoracic 'motors' flows close to hemolymph passing forward in the dorsal vessel, so heat cycles back to the thorax. Retained heat in the thorax helps thoracic muscles warm-up on cold mornings. Honeybees enhance heat exchange by having the aorta make tight spirals in passing through the narrow petiole thus increasing the area of the aorta contacting the backwards flow of hemolymph.

Now for increasing heat loss. Large naked carpenter bees can lose heat readily. Cuticles of head, thorax and abdomen are smooth and devoid of hairs. Smooth surfaces radiate heat. A radiator increases the surface area to transmit more heat to the environment by convection. A bee's body, filled with hemolymph has a high heat capacity. The open pump circulation spreads heat against the inner surfaces of thin cuticles. When carpenter bees fly at about twelve meters per second, the cooling rate of the head may be ten times greater than that of the thorax. A cooler head draws heat from the flight motors in the thorax as the dorsal vessel passes the hot hemolymph forward. Backward coursing hemolymph through the thorax transfers heat to the abdomen, where the thin walls radiate excess heat away into the air stream. In warmed temperature-controlled spaces, carpenter bees increase their flight speeds. When ambient temperatures are lower, they hover close to the ground conserving heat.

Evaporative Cooling

Honeybees on hot days regurgitate dilute nectar from their crops onto their heads and bodies. The bee's head is an excellent radiator, but the fluid from the crop, voided through the mouthparts, evaporates in the moving air to provide additional cooling. One disadvantage of evaporative cooling is that bees must replenish the nectar in their crops from nearby flowers (Ref: Heating and Cooling).

Turbulent Transport

Turbulence allows recursive regress of eddies into eddies over a wide range of scales. In our flying bee, turbulent air around body and wings mixes chaotically and draws excess heat from the circulation. Turbulent mixing is highly efficient. Nernst layers are small, and the external resistance to diffusion within turbulence is low. Periodic driving of turbulent streams, however, is necessary to maintain a turbulent system, as turbulence continually dissipates energy. Forward thrust from thoracic engines creates this turbulence.

Turbulent transport also underlies the high efficiencies of metabolic reactions. In mammalian systems, laminar flows in blood vessels (low Reynolds numbers) still produce almost turbulent mixing, when these flows combine to traverse the valves and chambers of the heart. Labeled red blood cells follow ergodic and unpredictable paths within blood vessels. Models suggest that after sufficiently long stroboscopic observation, the path of just one red blood cell cycling around and around the circulation, if given sufficient time, would fill the entire volume of all the vessels it travels through. In vertebrates, as opposed to micro-machines, this "circulation volume" would be the entire volume of the body. We know about circulation volumes from the exponential distributions underlying renal clearances. Fractal arrays and optimal configurations of tissues having extensive surfaces closely apposed to an internal transport system form a structural basis for turbulence.

Diffusion Again

Let's consider again in greater detail the irregular jiggling displacements of an uncharged particle floating in hemolymph or within a microdevice, as all those irregularly bouncing water molecules strike the particle. We can calculate the particle's average or root mean square displacements at any intervals we choose. Then we can compare our observations with theoretical calculations.

First some history. In 1905, Einstein showed how small water molecules could jiggle particles like pollen grains observable under a microscope. In his classic paper: 'On the motion of small particles suspended in a stationary liquid according to the molecular kinetic theory of heat,' Einstein used statistics to show that many molecular "beatings" combined to bounce larger particles around. For particles smaller than about twenty micrometers across, the impacts falling equally on all sides failed to average out, thus giving the particle a kick in some direction. As we know, no particle anticipates where it may be kicked to next.

Particles experience viscous drag. Drag depends on mass, how fast a particle moves and a coefficient describing the viscosity of the hemolymph. If our particle's a sphere, we can use Stokes Law. Our particle, therefore, also feels this rapidly fluctuating force that averages to zero over enough time. If we multiply our particle's equation of motion by its displaced distances averaged over time, and we apply the equipartition law, we get Einstein's equation for diffusion. This equation says explicitly what the root mean square displacement is in parameters we can measure. However, we must watch our particle for a long time compared to the shorter intervals between hits. Einstein and Smoluchowski asked: in a time interval how far does our particle move from where it starts?

One-Dimensional Random Walk

Think of it this way. We now confine our particle to wander back and forth along a linear path. A bell-shaped Gaussian distribution defines the probability of the particle being at some specified distance from the starting point after a certain number of steps. The more steps taken, the wider the curve. Indeed our expected average distance from the start is just the length of each step times the square root of the number of steps taken.

A particle being pounded on incessantly goes with the flow, so imagine a particle between two organs. Every minute with a fifty percent probability, $P = 1/2$, it jiggles either ten units towards 'The Thorax' or with probability of a third, $P = 1/3$, it drifts towards 'The Midgut' or it remains for the interval where it is with $P = 1/6$. Our particle takes a random walk in one dimension, and its movements with time form a finite Markov chain. Assume also the midgut and thorax hold the particle if it arrives there. Knowing the distance between the midgut and thorax and our particle's starting position, we ask what place is it likely to reach first, and how long will it take to get there.

If the midgut and thorax are fifty units apart, and our particle is originally twenty units from the thorax, we can label its potential stops as $E1$ to $E6$ with our two extremes being the midgut and thorax. We give $E4$ as the vector $x = (0, 0, 0, 1, 0, 0)$ in which the ith component of this vector is the probability that the particle is initially at Ei. Vectors $(0, 0, 1/2, 1/6, 1/3, 0)$ and $(0, 1/4, 1/6, 13/36, 1/9, 1/9)$ are our particle's positions after one minute and then two minutes elapse. Using a transition matrix, we can imagine its location after k minutes:

Transition Matrix

Let Pij be the probability the particle moves from Ei to Ej in one minute. Let's say $P23 = 1/2$ and $P24 = 0$. These are our transition probabilities, and a 6×6 matrix $P =$ is the transition matrix.

$$\begin{matrix} 1 & 0 & 0 & 0 & 0 & 0 \\ 1/2 & 1/6 & 1/3 & 0 & 0 & 0 \\ 0 & 1/2 & 1/6 & 1/3 & 0 & 0 \\ 0 & 0 & 1/2 & 1/6 & 1/3 & 0 \\ 0 & 0 & 1/2 & 1/6 & 1/3 & 0 \\ 0 & 0 & 0 & 0 & 0 & 1 \end{matrix}$$

Each entry in our matrix is non-negative, and each row straight across sums to one. If x is our initial row vector, then xP gives the probabilities for the particle's position after a minute and after k minutes, the vector xp^k gives it, so the ith component of xP^k gives the probability that the particle is at Ej after an elapse of k minutes. We reveal the Markov chain using an $n \times n$ transition matrix P, and a $1 \times n$ row vector x. Positions Ei are the states of the chain. Now how does the particle move from one state to another? It gets from $E4$ to $E1$ in three minutes and goes from $E4$ to $E6$ in two minutes, but it cannot go from $E1$ to $E4$, because once the particle touches either the midgut or thorax, it stays there.

So our problem is not one of all actual probabilities but one when these probabilities are not zero.

A Huge Simplification

We can represent any Markov chain using a directed graph or digraph. The vertices are the states, and the edges tell if we can move from one vertex to another vertex in one minute. We draw an edge only if the probability of traversing that edge is not zero. We can also construct the graph from the matrix if we replace each non-zero entry of the matrix with 1 or certainty. The graph shows we can move from vertex to vertex only if a path between these vertices exists, but, most importantly, the shortest edge gives the quickest possible time to make the transition. The edges are strong connections between the vertices. Edges tell us which vertices we return to over and over, and which ones we visit a few times and do not return to.

More formally: if we start at $E1$ and remain there, the probability of returning to $E1$ later is 1, and we call $E1$, our midgut source, a persistent state. States between source and sink are transient states. Digraphs avoid difficult calculations. We see that Ei persists if and only if a bidirected edge connects vertex i to vertex j. If a path exists from vertex i to vertex j but not back from vertex j to vertex i, then vertex i is transient. Like the midgut source and the thorax sink, any vertex from which we cannot get to another state is an absorbing state.

Brownian Particle

Now to a real Brownian particle and Einstein and Smoluchowski's solution. Divide the time we watch the particle into intervals of hundredths of a second. Now after one short interval the particle moves to one place, and in the next short interval it moves again and so on. A single water molecule receives about 10^{14} strikes each second, so in a hundredth of a second, our particle receives 10^{12} hits. Collisions are random, so each step a particle takes is independent from the step it just took and the step it will take.

Where is the Particle?

After a long time can we locate our particle? No. We cannot know where it is. How about on the average? On the average how far has our particle progressed from where it started? Remember, the mean square of the distance traveled in the sequence of random steps is the sum of the separate distances. So if vector N is the vector distance from our origin after N steps, the mean square of the distance is proportional to N, the number of steps.

But is the distance proportional to the time? If this were true, the particle would have to have progressed at a uniform velocity. So each hundredth of a second the particle makes headway but only so much headway that our particle's mean square distance is proportional to time taken.

What About Dimension?

What happens in two dimensions? After all, we have two dimensions in our smallest hemocoel. Step randomly to the north, south, east or west each time. Once the walker returns to the origin, it starts over again, and then given enough time, there is a second return to the origin, and then a third return, and so on. Two-dimensional random walks, then, visit all points of the surface if walkers have sufficient time to complete their walks.

Now what happens to walkers in three dimensions? Here things are different. Very different. Walkers can go up and down as well as in the four compass directions. Use a standard six-sided die to determine your particle's movements. Now however, even if our walker takes infinite time and infinitely many steps, its probability of getting back to the origin or any other specified place in the hemocoel is only about a third or point three four. There is so much space available to get lost in, that unless our particle happens to make it back to the origin in just a few steps, it is most likely to get lost forever. There are so many ways for aimlessly wandering molecules to get lost. It is for this reason that three-dimensional hemocoels do not transfer particles from one position to another unless convection adds direction to diffusion (Ref: Walking Particles).

Forces On Our Particle

Suppose a convective force exists on the particle, not just a Brownian force. First our particles own inertia would slow it. Let some number be the coefficient of inertia or the effective mass of our particle. It is not our particle's actual mass, because our convective force pushes water aside as it pushes our particle. A unidirectional force gives a mass moving over distance. If the force is steady, there is a fluid drag proportional to our particle's velocity. The complex viscosity of hemolymph also resists this flow. We also irreversibly lose heat due to friction. We cannot get kT if we cannot have these heat producing losses due to drag.

Least Action

Suppose our particle moves from its original place to another place in an interval. Then the particle does it again but follows a different path through the hemolymph, but it gets to the same place it went the first time in the same amount of time. If we calculate the particle's kinetic energy at every moment along the path and subtract the potential energy at every moment and integrate it over the time the particle needs for the whole trip, the number we get is larger than the particle's actual motion.

The principle of least action states that the average kinetic energy minus the average potential energy stays as small as possible while going from one point to another. The true shortest path is the path for which this integral is least.

Why is this possible? Because were our particle to take any other path than the one it takes, its velocities would sometimes exceed and sometimes be less than the average velocity. Average speed through the hemolymph is the total distance traversed over the time. This means that the mean square of something varying around an average exceeds the square of the mean, so that now the integral for the kinetic energy would be greater if the velocity were irregular than if the velocity were uniform. This is another way of saying that our integral is minimal if the velocity remains constant, and we have just a uniform push and no random forces. Such solutions are always balances between holding on to the most potential energy while expending the least extra kinetic energy to keep the difference between kinetic energy minus potential energy as small as possible.

Action

Kinetic energy minus potential energy integrated over time is, of course, our particle's action. For each path there is a minimum action. Think of a circle as a locus of all points at a constant distance from a fixed point or as a curve of a specified length enclosing the largest area. Any shape for the perimeter other than a circle must enclose a smaller surface area.

So in considering paths through the hemocoel from one point to another, we imagine there is one true shortest path where the action is minimal, so that taking any other curve means taking a false path, a more energetically wasteful path, because if we calculate the action over the false path, the action is greater than if we took the shorter path. Or let's assume the particle takes a minimum path to start with. If we deviate from it in the first order, the function deviates from its minimum only by the second order. At any place along the curve, if we move a small distance away, the value of the function changes also to the first order but at a minimum. Taking just a tiny step to the side makes no difference at all in the first approximation. If there is a change in the first order when a particle deviates, the change in the action is proportional to this deviation. Reversing the sign of the deviation makes the action less so. At this point we can have the action increase going one way and have it decrease going the other.

It turns out that the path having the least action is the path satisfying Newton's law. However, in the hemocoel, we cannot forget friction. The principle of least action only works for conservative systems in which we obtain all our forces from a single potential function. But at the microscopic level of organization there are no conservative functions. The Lagrangian is the function integrated over time to give the action. The Lagrangian is a function only of the velocities and positions of particles.

Minimum Along the Entire Length

Consider an actual path of a particle through the hemolymph and say the true path traverses point A and point B. If the action is minimal along the entire path, then the path between a smaller section of the line, between points A and B for example, is also minimal. So every subsection of the path our particle follows must also be a minimum no matter how short we make our subsection. So now the statement about what happens over the entire longer path also describes what happens over any shorter piece of the path.

Lévy Flights and Superdiffusion

Now consider a particle walking randomly in a hemocoel being jiggled by the convective circulation of hemolymph as the hemocoel itself jiggles during flight. Our particle starts at one position, and takes steps in random directions. As we know, random walks may occur in all three dimensions depending on local conditions.

In some cases the jiggling may make the steps randomly long as well. Now suppose a 'walker' particle diffusing in the hemocoel jiggles as our bee flies. Now perhaps motion from the jigging combines with convection from normal circulation to add random velocity component vectors that force the random walker to pause for random amounts of time in between steps.

Normal Random Walk

In the hemocoel perhaps ‘most’ random walks spread out diffusing normally, the variance of a group of particles growing linearly over time. This variance describes the size of a typical group of diffusing particles. (Remember: average of the squares of the distance a random walker moves minus the square of the average of the distance the random walker moves.) Our diffusion constant is the rate at which the variance grows. Remember *D*, the diffusion constant, is large when particles move faster in water and smaller when they move slower in syrup.

For cases where a particle’s step length is random, *D* depends on the average *squared* step length rather than the length of an average step. Also, if the random walker takes one step every two seconds, it is reasonable to guess that *D* might be smaller.

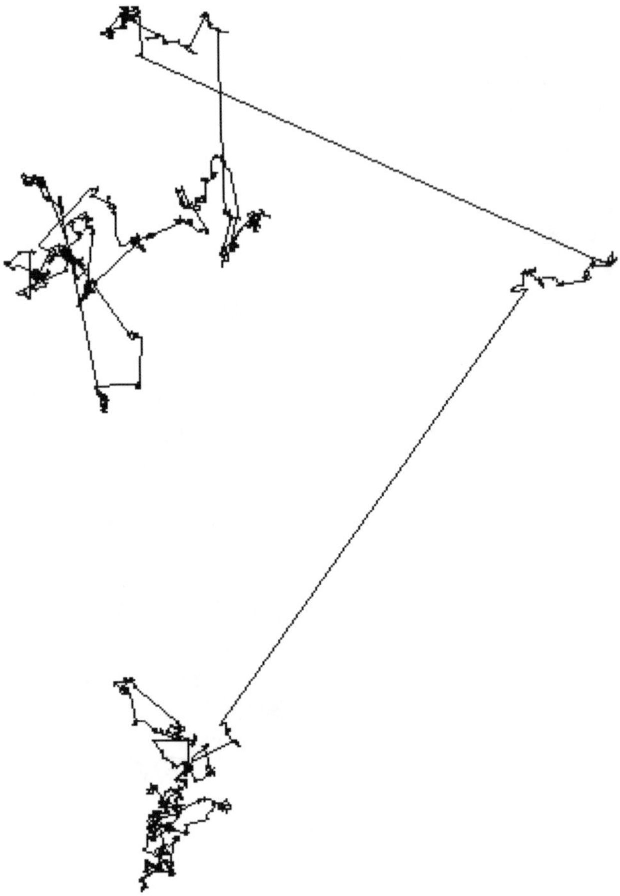

Lévy flight

Now what happens when the average squared step is very large compared with the usual step length or, we might say, infinite? Such a walk with very large segments are Lévy flights, and D becomes infinite.

The two pictures compare a normal random walk and a Lévy flight. The thousand steps adding up to each particle's trajectory are of random lengths. In the normal walk the probability of a long step is proportional to $L^{(-3.8)}$. In the Lévy flight, the probability of a long step is proportional to $L^{(-2.2)}$, which is more probable than the normal case.

Follow a normal random walk for a very long time to begin seeing 'normal' behavior, that is we can no longer make out the small steps. However, the average lengths of all the steps taken together determine the trajectory's pattern. In a Lévy flight, convection and jiggling create the long, infrequent steps, the flights. These long straight segments determine the trajectory's pattern. In a Lévy flight, the few, long rare steps, the flights, mostly determine a walker's position. Thus, in a Lévy flight, the individual steps do not average out.

Review of Lévy Flights

If Lévy flights occur, the variance increases faster than just linearly with time. The variance denotes the typical size of a group of random walkers, and is the average of the squares of the distance moved minus square of the average of the distance moved. D is the rate at which the variance grows. Diffusion is faster in watery hemolymph (large diffusion constant) than in hemolymph of high osmolality (small diffusion constant). Volume changes in the hemocoel complicate these relationships.

If variance equals time raised to some exponent, the exponent being one for normal diffusion, this exponent is larger than one, but usually less than two, during a Lévy flight. Superdiffusion is when the exponent is greater than one but less than two. Having an exponent of two would mean that all random walkers were moving apart from each other at a constant rate as might occur in an explosion. We still must learn what role, if any, Lévy flight dynamics and superdiffusion play in the bee's hemocoel (Ref: Lévy Flight Dynamics).

Conclusions: Generalized Cavity Transport

Generalized transport in a hemocoel employs diffusion and convective dispersal within a confined space. Convective forces from the open circulation and the jerky movements of walking and flying disperse the particles perhaps adding components of superdiffusion.

Hemocoels avoid the higher energies and limits imposed by a closed circulation and pump to maintain high pressure to force fluid through peripheral resistances. Hemocoel transport is miniaturizable while pump-tube transport is not. Hemocoel transport should be useful in microfluidic devices.

We may also model hemocoel transport using mass action and percolation theory. We can simulate diffusion and dispersion on an incomplete lattice. We can explore the heterogeneities of surfaces bordering on the hemocoel cavity to determine what exogenous factors facilitate or impede mixing.

If diffusion distances are short and movements minimal, might we even imagine patterns of waves traveling through the volume? As movements increase, do contact of particles with walls and organs increase or decrease as the cohesive and adhesive forces change? How might the densities, concentrations and other intrinsic properties of particles influence the dynamics of hemocoels (Ref: Jacobs and Hendrickson, 1997)?

Chapter ten

CONTROL

Overview

Homeostasis, or staying the same, is a typical property of open complex systems. Think of a cell, a bee and a law firm. Such systems react to disturbances by modifying themselves to oppose the forces creating the disturbances. Homeostasis rigorously controls many interdependent regulatory mechanisms to maintain current structure and function. But for complex systems as complex as bees and people, endurance is not enough; they must adapt to changes of the environment and must evolve. Otherwise, entropy will disorganize and destroy them.

Control systems maintain the internal state of a bee or device so that smaller systems depending upon the internal state continue to work optimally. A bee maintains her temperature as well as specific ions and molecules in her hemolymph. Controllers also permit her to adjust her response to changes in her environment. Threaten her, and she prepares to fly and sting.

After introducing control systems, we view molting as an example of system-wide control. We then see how controllers and controlled systems interact continuously and evolve.

Stability and Change

A bee is a hierarchical array of subsystems regulated by a web of control systems. The control systems together adjust the flow of energy and materials through the body maintaining a constant internal environment despite changes in the external environment (homeostasis). How is control maintained?

For a bee to live she must adapt to stress (trauma, infection, drying, increased temperature) and continue to function. Analysis of where the weak links are in her web of control systems suggests that her life depends on a few key control systems. We limit discussion to control of the hemocoel that entails coordinating many interrelated functions. As the volume of hemolymph fluctuates, a bee must alter the chemical content of her fluids including the proportions of cells, metabolites, ions, water and nitrogenous wastes, as well monitor uptake and discharge by her cells from and to the hemolymph.

Definitions of Control

Most control systems employ negative feedback, but some employ positive feedback. Negative feedback is operating when disturbances to a system evoke compensatory changes tending to diminish the effects of the disturbances returning the system towards its normal state. Positive feedback, unlike negative feedback, is explosive, and variables rapidly achieve a maximum or a minimum state. Any control system's adjustability is limited, and each control system operates best only within a restricted range. Within any range of control a disturbance may exceed the limits within which the control system can compensate for the disturbance. Failure to correct a disturbance may exert a domino effect, spreading rapidly and if related control systems cannot compensate for the failed control system, death ensues.

Feedback and Bee Stings

Negative feedback opposes change and restores stability. Positive feedback amplifies a disturbance's effects. Both types of feedback control a disturbance and may occur within the same system. A predator receives a bee-sting. If he moves away from the hive, it is negative feedback. But if the disturbance continues, angry bees may entice others to sting, until the entire colony emerges stinging. If the predator and the angry bees all die, together they have displayed positive feedback.

Feed-forward Loops

Positive feed-forward loops are inherently unstable, but feed-forward loops are common in endocrine and metabolic control systems. Positive feed-forward signals may increase flow through a metabolic sequence. Positive feedback occurs during blood clotting. The clotting sequence is inactive at rest, but once triggered, the cascade of clotting factors goes to completion, amplifying the original signal. The process self-limits as the clotting factors get used up.

If we disturb a system, and a variable remains constant, we expect to find negative feedback controlling that variable. If, on the other hand, a variable suddenly and rapidly increases or decreases, we expect to find positive feedback.

Redundancy

Many systems of receptors and effectors together control important variables such as body temperature or blood pressure. Having multiple controllers ensures that adequate control remains if some but not all controllers are disabled. Redundant control ensures stability.

Controllers and the Controlled

Controlling a variable means that a controller operates as part of a controlled system. What's controlled feeds back upon the controller but does not 'control' the controller. Feedback describes the moment-to-moment interaction between the controller and what's controlled. Feedback is continuous and changes over time as the system changes. Feedback filters through the controller's 'representation' of the interaction. Think of a fighter pilot watching the image of a laser bomb striking a target. The controller reads a composite image of its own actions superimposed upon an image of what's being controlled. The controller may even destroy what's controlled at the same instant as the system engenders a perception within the controller as when the pilot watches his bomb detonate on the cockpit screen.

To accomplish control, the controller changes some variables in the controlled system directly. At the same time, the controller senses the changes and feeds these back to what operates the controller. Each system's dynamics determine what the observed and controlled variables are. Uncontrollable disturbances may also influence the observed variables.

Set Point

Somewhere in the system resides the goal for the control system. This goal is the set point and is analogous to the set temperature for a thermostat. In the bomb analogy, the set point is the target's position. The controller observes the target, while continuously comparing its momentary representation with the internal set point of the target's coordinates. Seen from the controller, the loop begins with the controller's action that feeds back to create the controller's perception of what it just did. In a primitive control system no representation of the controlled system may exist distinct from the controlled system itself, as feedback flows directly from what's controlled back to the controller. When the controller's action on the system does not differ from the system's feedback to the controller, we have a harmonic oscillator.

Controllers respond to minimize the difference between what the controller observes and the set point. Analogously, a flying bee controls its temperature (Chapter 9).

Supercontroller

The idea of a super-controller overseeing hierarchies of lower subservient controllers exists in biology, cybernetics and the computer sciences. Teasing control systems apart from what they manage can reveal new patterns of control. Control of the hemocoel remains unstudied but probably contains useful secrets.

A hemocoel's local politics are unique. Control of the inputs and outputs of the hemocoel occurs simultaneously distributed over the entire space of the hemocoel along any open boundaries where cells contact hemolymph and entry and exit 'choices' occur. Individual entries and exits do not obstruct flow of hemolymph within the hemocoel. Compare a hemocoel with the filling and emptying of a football stadium as flowing crowds congest entries and corridors where each person's choice feeds back to change what the group does.

Seen from a higher perspective, sites of absorption and elimination from organs and cells are not the meta controllers for what enters and leaves each organ, but together, we might imagine a super controller directing all these sites, so that together orchestrated separate controllers might coordinate the hemocoel-system. Each cell and organ probably controls its own surfaces and receptors, individually responding to immediate local demands for metabolites and excretion without requesting outside information filtering down to each control point from a distant central processor.

Hemocoel as a Parallel Processor

The holy grail of computer processing is parallel processing in which many inexpensive individual microprocessors engage simultaneously. Supra-scalar processing resembles parallel processing, in that the hardware automatically locates instructions that launch at the same time. One popular aid is to place a cache or small amount of memory on each microprocessor. The cache holds only the most frequently called parts of the general program, so as to avoid energetically costly frequent retrievals from distant memory chips thereby enhancing overall speed. Because control of the hemocoel is diffused over so much volume and area, hemocoels remain active and robust even if point blockages develop or some controllers do not work.

Hemocoel Pipelines as a Microprocessor

We are tempted to consider the hemocoel as a microprocessor. Its unique parallel processing capability comes from the idea that the hemocoel 'pipelines' in that many steps of many parallel sequences operate concurrently. Designers increase a system's clock rate or speed by using scaling technology to make chips smaller and by reducing the numbers of levels of logic each cycle needs (more levels means longer time). However, using detailed wire and component models, many of today's designs scale poorly with technology (Ref: Agarwal *et al.*, 2000). If each step has limited time to execute, the time the hemocoel saves by pipelining is proportional to how many stages are active. Low-level "instructions" given to the "hardware of the hemocoel" might include something about the dynamics of the hemolymph. The number of stages of processing a hemocoel completes each second as in a processor would be the hemocoel's "clock rate". For example, in comparison, a personal computer using a 200-megaherz clock might execute two hundred million stages each second. We have no idea of what a hemocoel's clock rate might be.

The hemocoel is uniquely organized to incorporate 'suprascalar' tasks in which more than one set of instructions might be performed in separate places at each stage. Because the controlled units of biological systems are cells or tissues, we might imagine a cache of memory, perhaps analogous to extranuclear DNA, held and used at the site of the processor itself. The cache would hold only parts of a central genetic program that the system most frequently refers to, thereby avoiding having to call on more distant memory.

Safety Factors

Hemocoels have huge safety factors or loading tolerances. Hemocoel safety factors are much larger than those for pump-tube systems. Hemocoels work when full or almost empty. Were we to incorporate loading tolerances into our models, we might learn what redundancies and fail-safe mechanisms we might use to prevent failure of our smallest devices. For example, small machines have many places in them where a point defect can cause the entire machine to fail (Ref: Drexler, 1992). As in the human circulation, an embolus in a coronary or cerebral artery can spell disaster. This default assumption, however, does not usually apply to machines built on the macro-scale, because tolerances are larger. Many macro-scale machines continue to work despite numerous point defects. Because even a large number of point blockages fail to stem flow through a hemocoel, macro-machine assumptions might aid in modeling a small system analogously, so that its ultimate design might tolerate a high density of point defects. What should be important for hemocoel models would be the configurations of its surfaces as well as their changing fluid interfaces.

New Models for New Control Systems

Understanding hemocoel dynamics so as to be able to model a hemocoel might lend novel insight to creating potentially useful control systems for our smallest devices if we could reduce the number of centralized controllers and their connecting 'wires.' Remember: too much fluid within a pump-tube system, such as our heart and blood vessels, can cause pump failure (congestive heart failure) leading to overall system failure (death). One human remedy is to take a diuretic to eliminate the excess fluid and to increase cardiac function with digitalis, but during fluid overload or paucity, all organs of a hemocoel continue to function without any tampering from without. In insects especially, most adjustments are local. Many physiological regulators reside in close proximity to the functions they control so transport is by definition short. Short distances mean fewer 'wires,' shorter wires, and less weight.

Molting becomes now our example of system wide control during metamorphosis. As old cuticle degrades and new cuticle grows, the hemocoel is a reservoir for recycled parts and energy. Even though adult bees do not molt, many insects do, and in bees, cuticle renews just before metamorphosis when a bees passes from larva to pupa and from pupa to adult.

Metamorphosis

Pliability and thickness of the exoskeleton differ in each developmental stage. Bees, butterflies and moths undergo 'complete metamorphosis,' in which the egg becomes a larva that receives adequate nutrition for growth and then enters a resting stage, or pupa. A pupa's cuticle is soft, but the wax walls of a brood cell further protect it. The pupa does not move around; it does not feed, and it is extremely vulnerable to predators and parasites. Even though pupae appear quiescent, they reorganize themselves internally very rapidly. Wings form internally to appear externally just before the pupa changes to an adult.

More than ninety percent of insects display complete metamorphosis. Insects undergoing incomplete metamorphosis avoid the pupal stage. They shed their exoskeletons at intervals (ecdysis or molting) while moving and foraging. Growth follows each molt, as the immature form, called an instar, grows stepwise to adulthood. Instars may molt four to eight times, and in some species, molting occurs thirty times. Each time an instar molts, its body is usually pale and soft, but the body soon swells in an hour or two before the exoskeleton hardens and darkens.

Molting

Insect cuticle is tough. Cuticle resists trauma and repels water. Apodemes or attachment points inside rigid cuticles are firm supports for muscles that move the body. Sheets and tubes of cuticle form wings and appendages. Cuticle protects delicate internal parts from predators and parasites, making insects tougher prey, but because cuticle is hard and tight like a coat of armor, and because cuticle stretches minimally, cuticles restrict continuous growth. The principle structural component of cuticle is chitin, a modified cellulose molecule. Chitin and products from the old cuticle are absorbed and probably recycled as about eighty percent of the new cuticle may contain materials from the old (Ref: Chitin).

Growth and changes in an insect's form occur only after molting or during metamorphosis after new soft cuticle has replaced the old. Two hormones, juvenile hormone and molting hormones (ecdysteroids), control molting. Rising titers of these hormones in the hemolymph drive epidermal cells to synthesize DNA and RNA in preparation for molting. If juvenile hormone is present before a critical molt, the insect retains its larval characteristics. If juvenile hormone is lacking but molting hormones are present, the insect becomes adult.

Many factors regulate molting's complicated sequence. The sequence begins with turning on production of hormones that initiate molting. The same epidermal cells that grew the old cuticle let it slough in a controlled manner by first pushing the old cuticle outwards so it may be shed before the same cells deposit new, soft cuticle. Growth or expansion of the body occurs abruptly after the molt to be followed by hardening or sclerotization of the new cuticle.

The epidermis, a thin layer of cells lining the surface of the body below the cuticle, secretes the new cuticle. The cuticle varies in toughness and thickness in different regions of the body and at different stages. Before the old cuticle sheds, the epidermal cells secrete digestive enzymes into the developing space between the old and the new cuticles. Stereotyped movements, such as rotating the abdomen, free the old cuticle from the epidermis. Active transport of potassium ions and bulk flow of water into the space between the old and new cuticles produce a molting fluid that may buffer the pH changes created as the old cuticle is digested and broken down. The insect splits the old cuticle by increasing internal hydrostatic pressure and swelling its body. Some insects even swallow air filling their alimentary tracts displacing the hemolymph to expand the hemocoel that stretches the cuticle. Some insects may even pump hemolymph into the thorax expanding and stressing the old cuticle, splitting it along specific lines of weakness.

In both types of metamorphosis, as each stage grows, the cuticle around it gets too small like tight fitting trousers. This old cuticle finally splits to reveal a fully formed new cuticle around the new body. Molting ceases once an insect is fully grown. Materials and energy for renewal of the exoskeleton must all pass through the depot of the hemolymph, and recycling of metabolites and controllers within the hemocoel determines how materials distribute to the body.

The soft new cuticle covers a very soft body, so hemolymph probably serves as a hydrostatic skeleton during this vulnerable period. After the body expands and the cuticle hardens, the volume of hemolymph decreases.

Molting during metamorphosis is only one example of system wide coordination of myriad individual controllers that accomplishes a very complex task. But first, how do controllers and control systems interact?

Metasystem Transitions

Now consider a system S that might synthesize cuticle on all surfaces of a bee simultaneously. In different locations on the body, cookie-cuttered copies of this synthetic machinery exist with a few local variants to specify different densities and thicknesses of cuticle. These patches of cuticle-manufacturing machinery unite into a new system S′ having the S-type cuticle producing systems as its controlled subsystems. We posit a controller for the behavior and production of our subsystems. S′ is now a metasystem with respect to S, and to create S′ we perform a metasystem transition.

By creating a series of metasystem transitions, bees may construct a multileveled control system increasing the complexity of control at each level. Each metasystem transition produces a higher level of organization, the metalevel, in relation to the level of organization of the subsystems. In the most general terms, a single relatively autonomous system that produces cuticle integrates itself into a larger web of systems that control it. Similar to the cell-organism analogy, in progressing from fertilized egg to embryo to adult, primary control gradually shifts from a cell to the organism.

The classical example of transition to a metasystem would be the development of a bee. S is an egg cell, surviving initially on its own. Following divisions, cells aggregate together to create a larva, an entirely new entity. Initially control entails holding daughter cells together, but following successive metasystem transitions, as cells integrate into tissues, tissues into organs, organs into an adult, subsystems specialize to create a multileveled interwoven hierarchy of structures and functions, all controlled by overarching humoral, endocrine, tracheal and nervous systems.

To create a metasystem, we first duplicate a subsystem and establish control over its multiple copies. The original system S is the scope of the metasystem transition, and the number of integrated systems within S is the scale of the metasystem. The minimal scale of a metasystem transition or MST is one. In a control system with many levels, each level associates with functions characteristic for its level. Each time we make a metasystem transition, we generate a new 'super' level of control. If A is activity at the top level, then each new metasystem transition, creates an additional new level of control, A′. A′ controls what the A level does.

Are Super-compilation Programs Natural Controllers?

In a super-compilation computer program, a metasystem transition relieves us from thinking of each single program as running alone, to considering each program to be now an input program into a larger metaprogram. The metaprogram now oversees a process that executes a sub-program. The subprogram treats the free variables of the input program and their interdependencies as subjects for its own analysis, while at the same time, optimizes the sequences of code in the input program.

To accomplish these feats, one part of the super-compilation program constructs a large algorithmic tree from the parts of the input program, while another part of the super-compilation program analyzes this tree, recognizes any recurring patterns, then prunes it to reduce the size of the algorithm making the whole program more efficient. The super-compilation program finally merges nodes together and deletes redundant sub-trees.

Would study of a bee's hemocoel facilitate locating analogous natural algorithms of control? If yes, might we then be able to dissect these 'natural hierarchies' or algorithms to reveal more of how biological control systems function? (Ref: Super-compilation).

Network Flow Problems in the Hemocoel

The hemocoel model solves several general routing problems that are best visualized using graphs. Network problems have the following form: Given a graph, where each edge has a capacity c, a source vertex s and a sink vertex t, the problem is to find out what is the largest flow one can route from s to t while respecting the capacities of each edge along our flow's course. Such routing ideas serve not only in design of pump-tube vascular systems, plumbing, or conduits. Finding the most economical way to move particles, call them goods, through a set of points is to solve a network flow problem. The dynamics of a hemocoel solve routing problems, and similar problems plague distribution system's allocation of resources in communication networks and scheduling situations. We refine network flow problems if we consider them as patterns involving shortest paths and edge vertex connections.

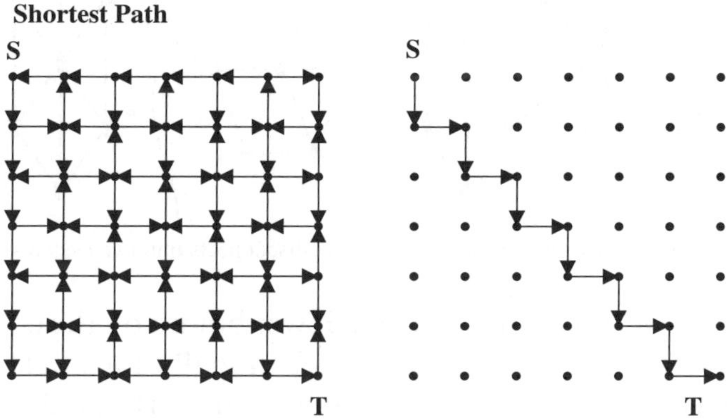

Let our graph be now an edge-weighted graph representing an adequately filled large hemocoel having a starting vertex S, for a substance and an end or recipient vertex T. We must find the shortest path from S to T. Again, a hemocoel solves this problem automatically.

Edge Vertex Connectivity

When volumes of hemolymph are low, the space of the hemocoel is confined to regions where there is moisture, so we encounter problems of edge vertex connectivity. Consider the hemocoel graph having an optimal throughput from S to T. What is the smallest subset of vertices or edges, that if severed, disconnect the hemocoel, or if we ask the question differently, what is the smallest subset of vertices or edges, that if they become non-functional, might isolate S from T?

Vertex connectivity may not be less than the combined connectivity of all edges, because if we delete one vertex incident on each edge of a cut, the graph disconnects, preventing flow of substances through it. Of course, smaller subsets of vertices might also work.

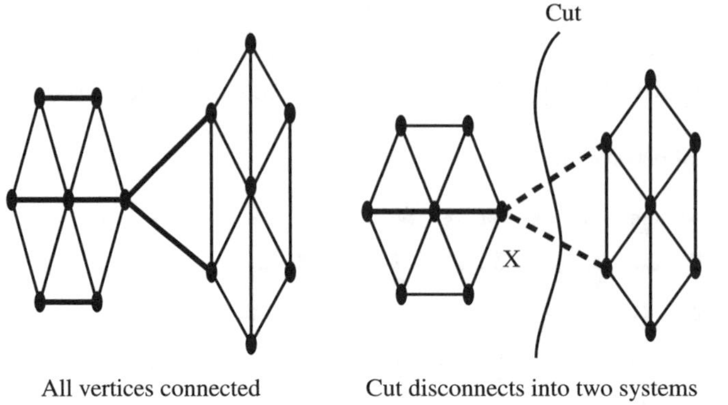

A minimum vertex degree places a lower bound on the connectivity of vertices and edges, because deleting all of a single vertex's neighbors (or the edges to all its neighboring vertices) disconnects the graph into one larger and one single-vertex portion. We may formulate many practical problems of linear programming as network flow problems indicating the power these models have. We can create special purpose network flow algorithms to solve such problems faster than we might with methods of general-purpose linear programming (Ref: Network Flow Theory).

Metasystem Transitions Create Modelable Complexity

A complex system is present if its global behaviors result from interaction of many small parts. The behavior emerges from the system as a whole and we cannot predict this behavior from just understanding the rules that govern the behavior each part individually. As seen in Chapter 2, transition from a simple to a complex system is neither simple nor discrete.

Complexity and emergence are only linguistic labels for diffuse problems in hemocoel dynamics, economics, artificial life, artificial intelligence, neuroscience, and even cultural change and development. Might we probe deeper to ask of any of these systems why 'something' we observed 'happened'? For our 'explanation' to be adequate at least at some level means we must know a sequence of events and interactions that led up to our 'something.' So we would be asking our old question again: can we 'explain' or infer connections between the initial state of our system and what changes to produce the 'something' we observe?

Asking these questions presupposes we understand on some acceptable level, the tools and techniques we use to define our system's rules, and we also must accept the validity of any theory that 'defines' our tools. But, still, our great open question always necessitates a leap of faith. The question still is: can we predict the outcome from an initial state without having to calculate every interaction?

This means, do we have or can we obtain a sufficiently deep enough understanding of the system that we can imagine some minimal number of symmetries for us to calculate the outcome? Or given the outcome, may we go stepwise in reverse order back to some space of initial states? If we can do this, we now have our mapping from the space of initial states to the space of outcomes. And we'd be almost home.

For if we can simulate the hemocoel system, and our model arrives at the result we expect, may we assume that the information hiding in the intermediate steps of our simulation explains the original sequence? Maybe yes, and maybe no. But if we can reproduce the behavior of the hemocoel and control it at each step, will we understand? Surely if we understand the system, we should not need to simulate its behavior. We would understand both the circumstances necessary for each step to the outcome. We would understand the correlation between the initial state and the outcome.

We have clearly then a continuum of levels of difficulty and a continuum in the complexity of our analysis. This is truly quite a deep issue. There is of course no discontinuous separation between emergence and non-emergence. Emergence then results from a 'phase change' in how much computation we must do to optimally predict outcomes. To imagine this scenario in computational terms, we must compute some minimal amount to predict the outcome.

Ultimately all useful predictive knowledge is in the accumulated interactions and the time required to complete our computations and thus depends on our machines and the time we need for computation. For finite computations, the time required using different Turing machines is related by an arbitrary polynomial. If this phase transition is real, it should not be machine-dependent. We measure the complexity of a step in terms of its Kolmogorov complexity in other words the length of its minimal description. The intuitive idea is that increases in Kolmogorov complexity often offset any decrease in computational steps (Ref: Modeling, Rent's Rule, and Kolmogorov Complexity).

So in an emergent system, our understanding can be at best zero. That we cannot predict emergent properties, stems not from any failure to understand, but from an inherent property of the system, brought about at least in part by the accumulation of interactions. So understanding this, we need no longer deal with any explicit dichotomy between emergent and non-emergent phenomena. Our perceived lack of understanding is really just another way of describing the complexity of the map between the initial state and our final phenomena. In the sense that lacking knowledge of initial conditions usually causes increasingly poor predictions is analogous to a discrete version of chaos. Any single phenomenon may fall anywhere in the spectrum between trivial prediction and emergence.

Remember chaos aids distribution because it aids transport. Transport limits turnover within bodies. Consider again how movement of virus or malarial parasites from a mosquito's gut to its salivary glands depends on transport within hemolymph. Also cooling in devices, bees or animals. Coolants moving within the interiors of systems to be maximally effective must penetrate into both superficial and deep compartments. The more intimately a coolant associates with the internal surfaces producing heat and the external surfaces radiating this heat, the more controlled becomes the transfer of heat (Ref: Chaos and Control).

Chaos works because particles, be they molecules or cells, suspended in blood or hemolymph responding to a chaotic component operating within their transport modality can explore a much wider range of values and potentially enter a wider range of spaces available to them within the body or device than could molecules or cells transported solely by rhythmical oscillations of their transport medium. Chaos introduces plasticity to cope with unpredictable changes in the environment. One direct way to follow particles through hemocoels is in (Ref: Localization Within Hemocoels).

Even though we cannot make an algorithm to optimize a general computer program, especially one containing chaotic elements, practical optimizations are possible because real biological and even computer programs contain redundancies that may lead to efficiencies. Might our local cuticle synthesizing 'programs' or machines optimizing their own code over geological time have figured out the best local way to perform their part in the synthetic function? Or simplistically, suppose the local machinery learns to be five times more efficient, and so the superprogram containing a metaprogram that optimizes and compliments local methods gets to be twenty times more efficient. So then together they then become multiplicative, speeding the process around a hundred times. I probably have to stop now and draw my conclusions.

Chapter eleven

GOALS AND CONCLUSIONS

Goals

We wish to correlate structures and functions at the nano- and micro-scales with what happens at other scales. We usually take three steps. First an idea surfaces that many feel is more fiction than science. Later, others realize the idea might be feasible, so they take it up. Lastly the developed idea enters research labs and university curricula.

Can We Make a Bee?

Eons of biological evolution have given us miniature devices we might copy. At the cellular level systems are collections of catalysts, sensors, skeletons, pumps and motors. Cellular machines create components that self-replicate, produce power, regulate its consumption, store information and maintain the internal environment. Analogous organizations and functions exist at the organ and organ system levels of complexity. At all levels complexity and emergence complicate our understanding. Complexity and emergence are properties that manifest themselves when any one formalism grows incapable of capturing all of a system's properties.

Before building a bee-like device, we must understand what the surfaces and interfaces of small hybrid structures can do. We need to develop capabilities across the range of complementary length scales between structures of tens of nanometers, like proteins,

DNA, and viruses joining our capabilities on the micron scale to our knowledge of how to manipulate cells and cellular assemblies.

We have had some successes. We can isolate whole systems, such as flagellar motors, and modify their biochemistry. Our understanding of microfluidics lets us handle and analyze minute quantities of liquids in the laboratory. Nanoprobes target cells; nanochips process DNA, and nanoscaled biochemistry labs on chips now depend on technologies that are approaching how integrated semiconductor devices transformed electronics and computation. We are learning to work small.

Scaling Holds Us Back

We use scaling laws to help us bridge between single molecules and higher order architectures, but we encounter problems with our 'understanding.' Integration is an emergent property of all natural systems including bees and micro-devices. Be they biological or engineered, systems are not mere collections of their parts. If they were, these systems would be subdivideable, and they are not. A system's ontology depends on context that in some complex way uniquely 'defines' a system's components. Outside 'the' system the components have different meanings. If we remove components from a system, the system looses its original identity and changes to something else.

Because systems 'emerge' we cannot predict when 'system properties' arise from assemblies of parts, but we must realize our inability to 'understand' is not from our failure to comprehend but from these inherent 'systemic properties' brought about in part as interactions accumulate.

Because dissecting cannot reveal how a system works, we must redefine our ideas of 'emergent system' and 'understanding' to include modeling as a new way to obtain information. If we can 'simulate' a system, and our model or simulation arrives at the result we expect from the original system, we might then say we 'understand' the correlation between the initial state and the outcome in both.

Conclusions

In this book I have spoken of several kinds of connections between the smallness of bees and insects compared with our smallest devices. I began with reasons to study insect circulations for their intrinsic interest and roles as possible models for devices as well as practically to understand their roles in the vector transmission of disease.

I showed how our closed pump-tube circulations don't work when reduced to insect size. I showed how open circulations differ from our own and how this difference permits open circulations to be miniaturized, some even capable of passing through a needle's eye. I then told of hemocoels and their graphs, and how shrinking hemocoels might improve their efficiency by shortening the distances hemolymph and particles must travel. Lastly, I spoke of diffusion on a surface and in a volume and how the probability of transmission by diffusion increased when the volume of hemolymph became compressed into a two-dimensional sheet.

It is time, therefore, to attempt a few generalizations, slight as the hard data may be. Accordingly, my first generalization is that insects can be models for our devices as they are all smaller and more efficient than our stand-alone devices thus far. If it is true that hemocoels really work better when shrunk, then insect 'smallness' is worthy of emulation.

Our second generalization based on a bit more hard data, but proposed in the same experimental way, is that shrinking decreases efficiency of a closed circulation but increases the efficiency of an open one.

Comparing Graphs

We compared the graphs of a closed tube circulation and an open cavity circulation. The graph of the closed circulation is a directed Eulerian trail. For each stroke of the heart-pump, blood flows in only one direction, from arteries and arterioles through capillaries to veins and back to the heart.

An open circulation or hemocoel exploits the principle that the shorter the path length, the faster, more direct and more economical is transmission through the system. Hemocoels remain the same size, but the volume of hemolymph in them and hence, the network graph for hemolymph within a hemocoel, grows and shrinks.

The graph of the hemocoel resembles a Moore graph in that every vertex potentially connects to every other vertex in the cavity directly. These connecting edges are identical and unweighted. Advantages of a cavity circulation are that nodes in the hemocoel are and remain dimensionless, featureless vertices. Transmission is potentially bidirectional. There are no hubs, as every point in the hemocoel potentially possesses direct 'airline' connectivity with every other point. Routes are lines or edges that are not vulnerable to point blockages as a fixed network of vessels would be. As an exponential network, the hemocoel is more stable than a scale-free network. Circulation is quite independent of the volume of hemolymph in the smallest hemocoels, and shortcuts through the two-dimensional film would still be possible.

Because each new edge forms or disappears independently from any other, the graph of the hemocoel does not form clusters, as neighboring vertices are now no more likely to be linked than would be any other randomly chosen vertices. Unlike a closed tubular circulation, nodes are eliminated in the graph of a hemocoel making the hemocoel system less vulnerable to point defects.

Randomly connected hemocoels differ from the connected lattices of a closed circulation. Introducing just a few 'shortcuts' reveals characteristic lengths that are closer to random graphs than to lattices. Adding a few shortcuts changes the hemocoel's dynamics appreciably, but adding more shortcuts is not better. Adding shortcuts means to a bee or device adding an energetic cost of volume and weight to be supported. That this may have been tried many times during evolution seems logical because evolution seems to have worked insects and circulations around to sit at the point where hemocoels are maximally functional.

Our third generalization is about modeling: every model contains an unintended intervention on the part of the model maker. One does not feel the need of so many reservations in the case of this principle. But in the sense of the first two, it is perhaps the one that matters least. It refers to our sense of the world. It also more importantly refers to a style of thought.

Models Provide Explanatory Power

Why choose one particular model of a system over numerous possible others? Everything else being equal, which of course it never really is, our chosen model should be the simplest model that agrees with observations. As an acceptable even a good model it should agree to a reasonable degree of accuracy with most of the experimental data. Here I emphasize 'most,' as the model also implicitly says we ought to continually monitor up-and-coming newer techniques.

So if our aesthetic reasons are strong enough, then it is only necessary for a model to agree with most of our observations. A model also should possess explanatory power. The model of the bee hemocoel is in a modeling relation to the bee system and by implication to those of other arthropods, and so it is not just a simulation.

The idea of explanatory power forces us see about how to validate a proposed model. A model is valid only if we can test it, and most tests of the hemocoel have yet to be devised. However, models are not merely explanations, but they are probes to provoke nature into showing us behavior that a trained observer can then accept as an observation. Observations provide further inductive evidence for confirmation or counterexamples. As Karl Popper said, "science is not to validate models but to falsify them." Models help us spell out testable observations and predictions. If our predictions fail, the model is falsified, and then we must reformulate or abandon the model. So I leave this last as a challenge to the reader. I have found no reason to abandon this model yet.

It is also time to attempt a few simplifications of the whole subject by way of summing it up and of coming to an end. With a few exceptions, all of the examples we have used have been pictorial. The image has been descriptive or explanatory of some image's subject. Said a bit differently, I have accompanied the stated thing with a restatement, and this restatement has illustrated and helped define the thing stated. The thing stated and its restatement together constitute an analogy.

The Value of Analogies

In conclusion let's talk briefly of the value of analogies. The fundamental ideas of the human spirit are vast collections of analogies that have helped make our thinking what it is. From Einstein's train all the way to Schrödinger's cat. The images inherent in these analogies include more than just figures of speech. Here we not only think these images but of the analogies of which our images are only parts. Analogies, most elusive, are always our larger subjects. These, our 'pictorializations,' are what stay in our heads to bias or spur how and what we think next. These images and the words that accompany them and the models resulting from them allow us to sometimes transcend our world while making our lives more interesting, exciting and livable. It is with this idea I close and hope it will supply some impetus and imaginative dynamism that should continue to constitute our science.

REFERENCES

NOTE

Because this book concerns rapidly evolving disciplines, I restrict this list to a few survey books and papers. Many possible illustrations have been omitted to save money since many become almost immediately dated, while others are so basic as to be in many references. In a few instances I repeat references in more than one chapter for those who read only one chapter. Current definitions, specific models, equations and techniques are available in multiple formats online; coverage ranges from general to deep.

Readers unacquainted with a specific literature such as insect morphology, microfluidic devices, graph theory and quantum dots let's say, might first search the categories and terms online progressing deeper as needed.

For example, two excellent mathematical sites are mathworld *http://mathworld.wolfram.com/about/mathworld.html* and The Math Forum Internet Mathematics Library *http://mathforum.org/library/toc.html* For those having a geometrical bent, the Geometry Junkyard *http://www.ics.uci.edu/~eppstein/junkyard/all.html* is a good starting place.

Equally general and deep websites exist for other subject areas.

GENERAL REFERENCES

Barenblatt, G. I. (1996). *Scaling, Self-similarity, and Intermediate Asymptotics.* Cambridge Texts in Applied Mathematics, Cambridge University Press, Cambridge. (A clear book about how scaling or power laws reveal a fundamental property of self-similarity in that self-similar phenomena repeat themselves in time and space. Self-similarity simplifies analyzing mathematical models. Scaling occurs as the initial or boundary conditions disappear, but the systems are not close to equilibrium, a situation termed intermediate asymptotics. The idea of a renormalization group is realized in many problems including theoretical biology and the mechanics of turbulence and fracturing.)

Colbert, E. H. and M. Morales (2001). *Evolution of the Vertebrates: A History of the Backboned Animals Through Time*, 5th edn. Wiley-Liss, New York. (A classic general text of vertebrate paleontology surveying the fossil record.)

Chapman, R. F. (1998). *The Insects: Structure and Function*, 4th edn. Cambridge University Press, Cambridge. (This is a good introduction to insect anatomy and some physiology with good illustrations and references.)

Drexler, K. E. (1992). *Nanosystems: Molecular Machinery, Manufacturing, and Computation.* John Wiley, New York. (The first book in the field using conventional physics and chemistry to introduce the basic concepts of power, strength, friction, wear, thermal noise and quantum uncertainty plaguing our smallest machines, devices and systems.)

Folkow, B. and E. Neil (1971). *Circulation.* Oxford University Press. London. (This is still the best book for the fundamentals. Concepts are not dated.)

Gershenfeld, N. (2000). *The Physics of Information Technology.* Cambridge University Press, Cambridge. (A technical discussion of the limits of devices for graduate readers in science and engineering to appreciate the implications of physical theory at the fuzzy interface between hardware and software.)

Grout, R. A. (ed.) (1963). *The Hive and the Honey Bee.* Dadant and Sons, Hamilton, Illinois.

Gupta, A. P. (1985). Cellular elements in the hemolymph. In: Kerkut, G. A. and L. I. Gilbert (eds.) *Comprehensive Insect Physiology, Biochemistry and Pharmacology*, Vol. 3. Pergamon, New York, pp. 401–451.

Hyman, L. H. and M. H. Wake (eds.) (1992). *Hyman's Comparative Vertebrate Anatomy*, 3rd edn. University of Chicago Press, Chicago, 758 pp.

Jones, J. C. (1997). *The Circulatory System of Insects*. Charles C. Thomas, Springfield, Illinois. (Covers the circulatory systems of over a hundred insect species including embryology, anatomy, physiology and reflexive bleeding.)

Kerkut, G. A. and L. I. Gilbert (eds.) (1986). *Comprehensive Insect Physiology, Biochemistry and Pharmacology.* Pergamon Press 13 volumes. New York. (An authoritative reference compiling work from international insect physiologists, biochemists and pharmacologists.)

Lawry, J. V. (2001). Insects separate diffusing particles in parallel. In: *2001 International Conference on Modeling and Simulation of Microsystems 2001*, Computational Publications, Boston, pp. 254–257. (This paper gave rise to this book.)

Meakin, P. (1998). *Fractals, Scaling and Growth Far From Equilibrium.* Cambridge Nonlinear Science Series, Cambridge University Press, Cambridge. (This book applies fractal geometry and scaling ideas to structures forming under non-equilibrium conditions and emphasizes computer simulations, experimental studies and theoretical advances.)

Mullins, D. E. (1985). Chemistry and physiology of the hemolymph. In: Kerkut, G. A. and L. I. Gilbert (eds.) *Comprehensive Insect Physiology, Biochemistry and Physiology*, Vol. 3. Pergamon, New York.

Newman, M. E. J., S. H. Strogatz and D. J. Watts (2001). Random graphs with arbitrary degree distributions and their applications. *Phys. Rev. E.* 64: 026118 [17 pp].

Ruppert, E. E., R. S. Fox and R. D. Barnes (2004). *Invertebrate Zoology: A Functional Evolutionary Approach*, 7th edn. Saunders College Publications, Philadelphia. (Organized by taxonomic group, this text discusses the biology and phylogeny of invertebrates and their anatomical plans and patterns of development from the Protozoa to the Arthropods. References.)

Snodgrass, R. E. (1935). *The Principles of Insect Morphology*. McGraw Hill, New York. (An oldie but goodie. Writing is good, vocabulary is explained, and drawings are excellent. Many of these illustrations in modified form are sprinkled throughout the insect literature. Is it time for a fresh look?)

Winston, M. I. (1987). *The Biology of the Honey Bee*. Harvard University Press, Cambridge, Massachusetts. (Reviews origins, evolution, anatomy, development, nutrition, behavior and some physiology with references.)

CHAPTER ONE: WHAT'S IN THIS BOOK

(Ref: Arthropod Vectors)

Deubel, V. and B. Murgue (2001). In: Service, M. W. (ed.) *The Encyclopedia of Arthropod-Transmitted Infections of Man and Domesticated Animals.* CAB International, Wallingford, UK, pp. 133–143.

Gubler, D. J. (1997). Dengue and dengue hemorrhagic fever: its history and resurgence as a global public health problem. In: Gubler, D. J. and G. Kuno. *Dengue and Dengue Hemorrhagic Fever.* CAB International, New York, pp. 1–22.

Christophers, S. R. (1960). *Aedes aegypti* (L.) *The Yellow Fever Mosquito: Its Life History, Bionomics and Structure.* Cambridge University Press, London, p. 738.

Holmes, E. C. and S. S. Twiddy (2003). The origin, emergence and evolutionary genetics of dengue virus. *Infect. Genet. Evol.* 23: 19–28.

CHAPTER TWO: DEVICES

(Ref: Dudley, 2000). *The Biomechanics of Insect Flight: Form, Function, Evolution.* Princeton University Press, Princeton. (Apart from being a comprehensive treatise on how insects fly, Dudley integrates the biomechanics of flight with the ecology and evolution of insects. References.)

(Ref: Feynman, 1945). 'Los Alamos From Below' from a talk given by Richard P. Feynman. Quote is from R. C. Tolman Professor of Theoretical Physics at Caltech, in the *First Annual Santa Barbara Lectures on Science and Society*, given at the University of California at Santa Barbara in 1975. The quote is from one of nine lectures 'Reminiscences of Los Alamos, 1943–1945.' L. Badash of UCSB is editing these lectures for publication.

(Ref: von Neumann, 1962). Reference is possibly from 'Method in the Physical Sciences' from *Collected Works by John von Neumann*, Vol. 6. Pergamon, Oxford.

(Ref: Microfluidic Chips)

Peterson, D. S., T. Rohr, F. Svec and J. M. J. Fréchet (2002). Enzymatic microreactor-on-a-chip: protein mapping using trypsin immobilized on porous polymer monoliths molded in channels of microfluidic devices. *Anal. Chem.* 74: 4081–4088.

F. Svec and J. M. J. Fréchet. Berkeley Labs Materials Science Division (*www.lbl.gov/.../ MSD-microfluidic-chip.html*).

(Ref: Limits)

Brodie, I. and J. J. Muray (1982). *The Physics of Microfabrication*. Plenum Press, New York.

Moore, G. (1979). VLSI: Some fundamental challenges. *IEEE Symp.* 16: 30.

Keynes, R. W. (1987). *The Physics of VLSI Systems*. Addison-Wesley, Reading, New York.

(Ref: Physics of Devices)

An excellent text on the physics of devices is Sze, S. M. (1981). *Physics of Semiconductor Devices,* 2nd edn. Wiley Inter-Science, New York.

An accessible introduction is Streetman, B. (1990). *Solid State Electronic Devices*. Prentice Hall, Englewood Cliffs.

Forster, F., R. Bardell, M. Afromowitz and N. Sharma (1995). Design, fabrication and testing of fixed-valve micropumps. *Proc. ASME Fluids Eng. Div. 1995 IMECE* 234: 39–44.

(Ref: Chitin)

Wainwright, S. A., W. D. Biggs, J. D. Currey and J. M. Gosline (1976). *Mechanical Design in Organisms.* Princeton University Press, Princeton, New Jersey.

Neville, A. C. (1993). *Biology of Fibrous Composites: Development Beyond the Cell Membrane.* Cambridge University Press, Cambridge.

(Ref: Self-reproducing Machines)

V. Zykov, E. Mytilinaios, B. Adams and H. Lipson (2005). Robotics: self-reproducing machines. *Nature* 435: 163–164. (Self-reproduction, central to life for evolutionary adaptation would be desirable in many engineered systems, but self-reproduction has yet to be incorporated in machines. These authors created simple machines or autonomous modular robots capable of physical self-reproduction using a set of cubes.)

CHAPTER THREE: BEAUTY BEFORE THE BEAST

(Ref: Power Laws)

Massive graphs have degree distributions that obey power laws. In a power law degree distribution, the fraction of vertices having degree d is proportional to $1/d^{\beta}$ for some constant $\beta > 0$. The starting point for these ideas is random graph theory and a model, introduced in 1959 by Paul Erdös and Alfred Rényi, in which when we add links or subtract links one link at a time, the pattern evolves into that of a classical random graph.

Erdös, P. and A. Rényi (1960). On the evolution of random graphs. *Mat. Kutato Int. Kozl* 5: 17–60. Newman, M. (2002). The structure and function of networks. *Comput. Phys. Commun.* 147: 40–45.

(Ref: Algorithm Design)

Kleinberg, J. and E. Tardos (2005). *Algorithm Design*. Addison-Wesley, New York. (An elementary book, teaching how we design algorithms starting with raw data and the roles of algorithms in computer science. Discusses implications and that local graph structure and long range connections in the graph give clues for finding paths through complicated networks.)

(Ref: Graph Theory)

Bollobas, B. (1998). *Modern Graph Theory*. Springer-Verlag, New York. (A classic book growing out of his *Graph Theory: An Introductory Course*, that includes the nature of the phase transition in the random graph process.)

Hierholzer, C. (1873). Über die Möglichkeit einen Linienzug ohne Wiederholung und ohne Unterbrechung zu umfahren. *Math. Ann.* 6: 30–42.

Skiena, S. (1990). Eulerian cycles. In: *Implementing Discrete Mathematics: Combinatorics and Graph Theory with Mathematica.* Addison-Wesley, Reading, Massachusetts, pp. 192–196.

Aiello, W., F. Chung and L. Linyuan (1999). A random graph model for massive graphs. In: *Proceedings of the 32nd Annual ACM Symposium on the Theory of Computing* (*http://math.ucsd.edu/~fan/random.ps*).

(Ref: Greedy Algorithm)

Zhang, Z., S. Schwartz, L. Wagner and W. Miller (2000). A greedy algorithm for aligning DNA sequences. *J. Comput. Biol.* 7: 203–214. (Discusses how a greedy algorithm can be faster than a traditional dynamic program for aligning DNA sequences differing only by errors of sequence into a new, theoretically optimal sequence.)

Greedy coins. Wagon, S. *http://library.wolfram.com/infocenter/MathSource/5187/*. This reference gives a simple, 'everyday' example of a greedy algorithm: 'A system of coins is called "nice" if it can represent any amount of money, and the greedy algorithm — repeatedly take the largest coin that will fit — for making change, always uses the smallest number of coins possible. The American system of coins — 1, 5, 10, 25, 50 — is nice since, for any number of cents A, the greedy algorithm for breaking A into a combination of coins will use the minimal number of coins. For example, a greedy breakdown of \$1.38 is 2 half-dollars, a quarter, a dime, and then 3 pennies, and \$1.38 cannot be obtained in fewer than 7 coins. Any total can be made by using only pennies.'

(Ref: Small World Networks)

Watts, D. J. (1999). *Small Worlds*. Princeton University Press, Princeton.

Watts, D. J. and S. H. Strogatz (1998). Collective dynamics of 'small world' networks. *Nature* 393: 440–442.

(Ref: Multiscale Modeling)

Lagana, K., G. Dubini, F. Migliavacca, R. Pietrabissa, G. Pennati, A.Veneziani and A. Quarteroni (2002). Multiscale modelling as a tool to prescribe realistic boundary conditions for the study of surgical procedures. *Biorheology* 39(3–4): 359–364. (These authors analyzed anatomically complex 3D models of vascular areas by invoking realistic boundary conditions to study local and global dynamics. Their finite element method simultaneously solved Navier-Stokes equations for local information and a nonlinear system of ordinary differential equations for global information. They first applied a finite element method using a boundary condition set of ordinary differential equations on two controls: a 3D model of a straight tube in a simple hydraulic network and a 3D model of a straight coronary vessel in a lumped-parameter model of the larger system. The results are very close to solutions obtained for networks of pipes.)

CHAPTER FOUR: YOU CAN'T SHRINK A WOMAN

(Ref: Hummingbird and Shrew Energetics)

Gass, C. L., M. T. Romich and R. K. Suarez (1999). Energetics of hummingbird foraging at low ambient temperature. *Can. J. Zool.* 77: 314–320.

Powers, D. R. (1991). Diurnal variation in mass, metabolic rate, and respiratory quotient in Anna's and Costa's hummingbirds. *Physiol. Zool.* 64: 850–870.

Peters, R. H. (1991). *The Ecological Implications of Body Size.* Cambridge University Press, Cambridge.

Krannenbarg, S., M. Muller, J. L. W. Gielen and J. H. G. Verhagen (2000). Physical constraints on body size in teleost embryos. *J. Theoret. Biol.* 204: 113–133.

Vornanen, M. (1992). Maximum heart rate of sorcine shrews: correlation with contractile properties and myosin composition. *Am. J. Physiol. Regul. Integr. Comp. Physiol.* 262: R842–R851 (0363-6119/92).

Suarez, R. K. (1992). Hummingbird flight: sustaining the highest mass-specific metabolic rates among vertebrates. *Experientia* 48: 565–570.

(Ref: Fractal Gels)

Sernetz, M. (1992). Fractal structure of gels determined by size exclusion chromatography or equilibrium techniques. In: *Proceedings of the IUPAC Symposium on Macromolecules.* Kahovec, J. (ed.) VSP, Zeist, pp. 423–429.

CHAPTER FIVE: BEE'S BODY

(Ref: Allometric Scaling)

West, G. B., J. H. Brown and B. J. Enquist (1997). A general model for the origin of allometric scaling laws in biology. *Science* 276: 122–127.

Kurz, H., K. Sandau, T. H. Dawson, J. H. Brown, B. J. Enquist and G. B. West (1998). Allometric scaling in biology. *Science* 281: 751a.

Damuth, J. D. (1998). Common rules for animals and plants. *Nature* 395: 115–116.

Banavar, J. R., A. Maritan and A. Rinaldo (1999). Size and form in efficient transportation networks. *Nature* 399: 130–132.

(Ref: Energy from Photosynthesis)

Rupa, D., P. J. Kiley, M. Segal, J. Norville, A. A. Yu, L. Wang, S. A. Trammel, L. E. Reddick, R. Kumar, F. Stellacci, N. Lebedev,

J. Schnur, B. D. Bruce, S. Zhang and M. Baldo (2004). Integration of photosynthetic protein molecular complexes in solid-state electronic devices. *Nano Lett.* 4(6): 1079–1083. (Abstract: Plants and photosynthetic bacteria contain protein-molecular complexes that harvest photons with nearly optimum quantum yield and an expected power conversion efficiency exceeding 20%. In this work, we demonstrate the integration of electrically active photosynthetic protein-molecular complexes in solid-state devices, realizing photodetectors and photovoltaic cells with internal quantum efficiencies of approximately 12%. Electronic integration of devices is achieved by self-assembling an oriented monolayer of photosynthetic complexes, stabilizing them with surfactant peptides, and then coating them with a protective organic semiconductor.)

(Ref: Flight)

Insect flight has a voluminous literature. Both Chapman (1998), in 'General References' and Dudley (2000) in the 'References to Chapter Two' are good introductions to this fascinating area.

(Ref: Respiration)

Flight is energetically costly. The thoracic muscles of flying insects have the highest known mass-specific rates of oxygen consumption of any muscles.

Spiracles: Harrison, J. F. (1997). Ventilatory mechanism and control in grasshoppers. *Am. Zool.* 37: 73–81.

Tracheoles: Weis-Fogh, T. (1964). Functional design of the tracheal system of flying insects as compared with the avian lung. *J. Exp. Biol.* 41: 207–227.

Enhanced ventilation from muscular contraction: Komai, Y. (1968). Augmented respiration in a flying insect. *J. Exp. Biol.* 201: 2359–2366.

CHAPTER SIX: CAVITY TRANSPORT

(Ref: Diffusion Rates)

Data from Macey, R. (1980). Mathematical models of membrane transport processes. In: Andreoli, T. *et al.* (eds.) *Membrane Physiology*. Plenum, New York, pp. 125–146.

(Ref: Trilobites)

Waggoner, B. M. (1996). Phylogenetic hypotheses of the relationships of arthropods to Precambrian and Cambrian problematic fossil taxa. *System. Biol.* 45(2): 190–222.

Clarkson, E. N. K. (1998). *Invertebrate Paleontology and Evolution*. Blackwell Science, Malden, Maine. (A good textbook with clear illustrations.)

(Ref: Negative Pressures in Hemolymph)

Brocher, F. (1931). Le mécanisme de la respiration et celui de la circulation du sang chez les insectes. Résultate de mas recherches pendant ces vingt dernières années. *Arch. Zool. Exp. Gen.* 74: 25–32.

(Ref: Random Graphs)

Newman, M. E. J., S. H. Strogatz and D. J. Watts (2001). Random graphs with arbitrary degree distributions and their applications. *Phys. Rev.* E64: 026118 (17 pp.) (These authors studied graphs having distributions of vertex degree that were significantly different from those having the usual Poisson degree distributions. They develop the theory of random graphs with arbitrary degree distributions and examine simple undirected, unipartite graphs and directed and bipartite graphs deriving exact expressions for the location of the phase transition where a giant component first forms, the mean component size, the size of the giant component, the mean number of vertices a certain distance away from a randomly chosen vertex, and the average vertex-vertex distance within the graph.)

(Ref: Routing)

Kleinberg, J. M., S. R. Kumar, S. Raghavan, S. Rajagopalan and A. Tompkins (1999). Trawling the web for emerging cyber communities. In: *Proceedings of the 8th World Wide Web Conference, Edinburgh, Scotland.* (The Web, a body of approximately 300 million pages, grows at about a million pages each day. That the set of Web pages lacks a unifying structure complicates implimenting a conventional database management and information retrieval system. Index-based engines searching the Web have created giant indices allowing users to retrieve the set of all Web pages containing a given word or string. Such search engines are, however, unsuited for deeper tasks because any topic typically contains several thousand or million relevant Web pages. How then might we design a search engine to retrieve from all these pages only the pages of most value to a particular user? How do bees extract specific molecules at separate locations from millions in the hemolymph?)

(Ref: Graph Models of Networks)

Many good references exist. Here are several:

Aiello, W., F. Chung and L. Lu (2000). A random graph model for massive graphs. STOC, Portland, Oregon. (This random graph model is a special case of a sparse random graph having given degree sequences involving log-size and log-log growth rates. For specified ranges they compute the expected distributions of the sizes of the connected components occurring with high probabilities.)

Albert, R., H. Jeong and A. Barabási (1999). Diameter of the World Wide Web. *Nature* 401: 130–139 (a seminal paper). Erdös, P. and A. Rényi (1961). On the strength of connectedness of random graphs. *Acta Math. Acad. Sci. Hungary* 12: 261–267.

Luczak, T. (1992). Sparse random graphs with a given degree sequence. In: *Random Graphs*, Vol. 2. Poznań (1989), pp. 165–182. Wiley, New York, U.S.A.

Barabási, A.-L. and R. Albert (1999). Emergence of scaling in random networks. *Science* 286: 509–512 (*http://www.nd.edu/~networks/science.pdf*).

Hofstad, R., G. Hooghiemstra and P. van Mieghem (1991). First passage percolation on the random graph. AMS. (Studies first passage percolation on a random graph having exponentially distributed weights on the links and for the complete graph described as a Markov chain in continuous time and recursive trees. The chain describes how many nodes can be reached in time *t* and the recursive trees describe a cluster's structure once it contains all the nodes of the complete graph.)

Jeong, H., B. Tombor, R. Albert, Z. N. Oltval and A. L. Barabasi (2000). The large-scale organization of metabolic networks. *Nature* 407: 651–654.

Albert, R., J. Hawoong and A. L. Barabasi (2000). Error and attack tolerance of complex networks. *Nature* 406: 378–382.

CHAPTER SEVEN: WHERE THE FLUID MEETS THE WALL

(Ref: Flow Equations)

A good mathematical introduction is Chorin, A. J. and J. E. Marsden (1993). *A Mathematical Introduction to Fluid Mechanics*, 3rd edn. Springer, New York.

A more biologically slanted introduction is Lighthill, J. (1975). *Mathematical Biofluid Dynamics*. Society Industrial Applied Mathematics, Philadelphia. Lighthill discusses the swimming of fish and bird and insect flight (high Reynolds numbers) and ciliary and flagellar propulsion (low Reynolds numbers) as well as respiratory flows, pulse propagation, blood flows, arterial disease, and microcirculation.

Lighthill, J. (1978). *Waves in Fluids*. Cambridge Mathematical Library, Cambridge. The useful, fundamental principles of wave motion in liquids and gases. (Lighthill was a polymath and a skilled teacher. His collaborations with zoologists began new fields. With Torkel Weis-Fogh, who succeeded James Gray at Cambridge, Lighthill showed how small hovering insects generate lift: the clap–fling–sweep sequence by which an essentially inviscid mechanism generates circulation round each wing, and thus lift. From Lighthill and Weis-Fogh we learned that the chalcid wasp, *Encarsia formosa*, achieves a lift coefficient greater than that of any man-made flying machine!)

Marsden, A. J. (1981). *Lectures on Geometric Methods in Mathematical Physics*. CBMS-NSF Regional Conference Series in Applied Mathematics 37. (This monograph explores symmetry, bifurcation and Hamiltonian systems in several diverse applications.)

(Ref: Yagerfaculty) *http://faculty.washington.edu/yagerp/microfluidicstutorial/tutorialhome.htm* (An online 'microfluidic tutorial' about work in the Yager laboratory at the University of Washingon. A situation somewhat similar to H-filters may be at work in hemocoels.)

(Ref: Minimal Surfaces)

Tuzhilin, A. A. and A. T. Formenko (1986). Multivalued mappings, minimal surfaces and soap films. *Vestnik Moskov. Univ. Ser. Mat.* No. 3, 3–12 (English Translation in Moscow University Mathematics Bulletin 41, No. 3.

Morgan, F. (1981). A smooth curve in R^3 bounding a continuum of minimal manifolds. *Arch. Rational Mech. Anal.* 75: 791–809.

Formenko, A. T. and A. A. Tuzhilin (1991). *Elements of the Geometry and Topology of Minimal Surfaces in Three-Dimensional Space*. American Mathematical Society Translations of Mathematical Monographs, Vol. 93.

Schoen, R. (1983). Uniqueness, symmetry, and embededness of minimal surfaces. *J. Diff. Geom.* 18: 791–809.

(Ref: Hysteresis)

Mammalian respiration including hysteresis: West, J. B. (1990). *Respiratory Physiology: The Essentials*, 4th edn. Williams & Wilkins (ISBN 0-683-08942-0).

Nunn, J. F. (1993). *Nunn's Applied Respiratory Physiology*, 4th edn. Butterworth Heinemann, Reed Elsevier, London (ISBN 0-7506-1336-X.)

(Ref: Lawry)

Lawry, J. V. (1966). Burrow irrigation in the echiuroid worm, *Urechis caupo*. *J. Exp. Biol.* 145: 343–356.

Lawry, J. V. (1970). Mechanisms of locomotion in the polychaete *Harmothöe imbricata* (L). *Comp. Biochem. Physiol.* 37: 167–180.

Lawry, J. V. (1971). The parapodial and segmental musculature of *Harmothöe imbricata* (L.) *J. Morph.* 135: 259–272. (Soft-bodied worms and caterpillars contain fluid surrounded by walls of muscle and hold their shapes by contracting against the incompressible fluid. Changes in volume are regional; contractions in the tail end increase the volume in the relaxed head end. High pressures create circular profiles in cross-sections of these bodies. We may analogize these forms using systems of springs.)

(Ref: Capillary Hydrodynamics)

Krotov, V. V. and A. I. Rusanov (1999). *Physiochemical Hydrodynamics of Capillary Systems.* Imperial College Press, London. (Book covers surface layers, drops and films and formulates two-dimensional and three-dimensional hydrodynamics for flows highly determined by surface phenomena. A must for hemocoel enthusiasts.)

(Ref: Svec and Fréchet) F. Svec and Fréchet Berkeley Labs Materials Science Division (*www.lbl.gov/Science-Articles/Archive/MSD-microfluidic-chip.html*)

CHAPTER EIGHT: SHRINKING

(Ref: Size Relationships)

Haldane, J. B. S. (1928). On being the right size. In: Gross, J. (ed.) *The Oxford Book of Essays*. Oxford University Press, London.

McGowan, C. (1994). *Diatoms to Dinosaurs*: *The Size and Scale of Living Things*. Island Press, Washington, D.C.

Moore, G. (1979). VLSI: Some Fundamental Challenges. *IEEE Spectrum* 16: 30.

Price, P. W. (1984). The world of the insect: size and scaling in moderately small organisms. In: *Insect Ecology*, 2nd edn. John Wiley, New York, Chap. 2.

Schmidt-Nielsen, K. (1972). Body size and the problems of scaling. In: *How Animals Work.* Cambridge University Press, London, Chap. 6.

Thompson, D. W. (1942). On magnitude. In: *On Growth and Form*. (See below.)

Wainwright, S. A., W. D. Biggs, J. D. Currey and J. M. Gosline (1982). *Mechanical Design in Organisms*. Princeton University Press, Princeton.

(Ref: Thompson, 1992) Thompson, D. W. (1992). *On Growth and Form*: *The Complete Revised Edition*. Dover, London. (Thompson wrote this classic of biology and science in 1917 and revised it in 1942 in which he develops his theory of transformation and interrelates the processes creating growth and form using physics and mathematics showing that not only do mathematical descriptions of forms and processes add precision to verbal descriptions, but that they also lead to Plato's idea that the *Book of Nature* is written in Geometry. This is a delightfully erudite book filled with interesting ideas and history.)

(Ref: Scale-Free Topologies and Robust Networks)

Aldana, M. (2003). Boolian dynamics of networks with scale free topology. *Physica D* 185: 45–66. (In a scale-free network, the proportion of nodes $P(k)$ having k links decays as a power law whose exponent is about 2–3. Existence of a phase transition occurs for values of scale-free exponent in the open interval 2–2.5. Fine-tuning that is usually required for stability in Boolian networks having random topology is unnecessary if the topology of the network is scale free.)

Robust Networks: Aldana, M. and P. Cluzel (2003). A natural class of robust networks. *PNAS* 100(15): 8710–8714. (These authors present a prototype for studying dynamical systems predicting the robustness of networks against varying parameters, demonstrating that dynamical robustness of complex networks is a consequence of their scale-free topology and in contrast, networks having homogeneous random topologies require fine-tuning of their internal parameters to sustain dynamical activity that remains stable.)

Pesin, Y. B. (1997). *Dimension Theory in Dynamical Systems: Contemporary Views and Applications.* Chicago Lectures in Mathematics, University of Chicago Press, Chicago. (This book is neither about dimensional theory or dynamical systems theory but covers subjects at the intersection of these two fields formulated within the last fifteen years.)

Baddi, R. and A. Politi (1997). *Complexity: Hierarchical Structures and Scaling in Physics.* Cambridge University Press, Cambridge. (This book provides a comprehensive discussion of complexity in physical, chemical and biological systems as well as in mathematical models to emphasize their common features by employing a uniform mathematical description with many examples.)

(Ref: Percolation)

Broadbent, S. R. and J. M. Hammersley (1957). Percolation processes I. Crystals and mazes. *Proc. Cambridge Philos. Soc.* 53: 629–641.

Grimmett, G. (1991). *Percolation*, 2nd edn. Springer, Berlin. (A comprehensive well-written account. References.)

(Ref: Self-organized Criticality)

Jensen, H. J. (1998). *Self-organized Criticality: Emergent Complex Behavior in Physical and Biological Systems.* Cambridge Lecture Notes in Physics, Cambridge. (Jensen discusses the spontaneous development of complex behavior in certain many-body systems whose dynamics vary abruptly and explains what constitutes SOC, where it occurs and then analyzes computer models and their mathematics.)

Pietronero, L., A. Vespignani and S. Zapperi (1994). Renormalization scheme for self-organized criticality in sandpile models. *Phys. Rev. Lett.* 72: 1690–1693. (These authors introduce a scheme characterizing the critical state and the scale-invariant dynamics of models of sandpiles. The results agree well with computer simulations and can be extended to situations having non-equilibrium stationary states that might apply to hemocoels.)

(Ref: Microfluidics and Drops)

Beebe, D., G. Mensing and G. Walker (2002). Physics and applications of microfluidics in biology. *Annu. Rev. Biomed. Eng.* 4: 261–286.

Cohen, M., P. Brenner, J. Eggers and S. R. Nagel (1999). Two fluid drop snap-off problem: experiments and theory. *I. Phys. Rev. Lett.* 83: 1147.

Deegan, R. D., O. Bakajin, T. F. Dupont, G. Huber, S. R. Nagel and T. A. Witten (2000). Contact line deposits in an evaporating drop. *Phys. Rev. E* 62: 756.

CHAPTER NINE: CHANCY TRANSPORT

(Ref: Diffusion)

A text presenting diffusion as a challenging physical problem in biological contexts: Okubo, A. (1980). *Diffusion and Ecological Problems: Mathematical Models.* Springer-Verlag, New York.

The classic text is Crank, J. (1975). *The Mathematics of Diffusion*, 2nd edn. Oxford Science Publications, Clarendon Press, Oxford. (This book is well written and starts with the basics describing the mathematical solutions of the differential equations of diffusion. Fick, in 1855, realized that transfer of heat was also due to random molecular motion, and he quantified diffusion by adopting the mathematical equation of heat conduction, so that the theory of diffusion still rests on the rate of transfer through isotropic substances. We naively assume hemolymph to be isotropic, so we can suppose the rate of transfer of a diffusing substance through the unit area of a section through hemolymph is proportional to the concentration gradient normal to the section.)

(Ref: Hardt, 1980) Transit times. In: Segal, L. A. (1980). *Mathematical Models in Molecular and Cellular Biology.* Cambridge University Press, Cambridge, England.

(Ref: Fractal Difficulties)

Meakin (1998) in our general list describes how complex, disorderly patterns develop under conditions far from equilibrium. He applies concepts of fractal geometry and scaling to the quantitative description and understanding of self-similar fractals, multi-fractals and examples that facilitate applications. Meakin emphasizes computer simulations and experimental studies and includes theoretical advances in diffusion-limited growth and the evolution of rough surfaces.

Colinet, P., J. C. Legros and M. G. Velarde (2005). *Nonlinear Dynamics of Surface-Tension Driven Instabilities.* Wiley Verlag, Berlin. (This book is an exhaustive monograph but a useful reference for experts describing thermal convections within fluid layers and the basics of instabilities driven by surface tension. The book emphasizes the generality of the mathematics and physics underlying results and methods and gives detailed derivations and analyses of nonlinear but non-dissipative structures having hydrodynamic instabilities in systems of interfaces. Definitely not a beginner's book but useful for modelers.)

(Ref: Gel-Filled Cavities)

Strathmann, R. R. (1989). Existence and functions of a gel-filled primary body cavity in development of echinoderms and hemichordates. *Biol. Bull.* 176: 25–31. (An extensive gelatinous material fills the cavities of deuterostome larvae and may be an elastic element opposing muscle pull permitting large larvae to maintain optimal shapes while minimizing skeletal supports. How fluid and diffusing substances traverse the gel is unknown.)

(Ref: Lungs and Struts)

Lawry, J. V. (1999). A hydrostatic Michell framework supports frog lungs. *Bull. Math. Biol.* 61: 683–700. (A braced framework of tubular struts in the walls and air spaces of frog lungs holds the lung open at zero transmural pressure. An orthogonal arrangement of the struts strengthens the framework and minimizes its weight; orthogonality is maintained as the lung inflates and deflates.)

(Ref: Mechanical Dispersion)

Koch, D. L. and J. Brady (1985). Dispersions in fixed beds. *J. Fluid Mech.* 154: 399–427.

A detailed discussion of diffusion, its connections to random walks, and the effects of interactions between the diffusing particles are in: Crank J. and R. Getz (1988). *A Primer on Diffusion Problems*. Wiley, New York.

Oxaal, U., G. Flekkøy and J. Feder (1994). Irreversible dispersion at a stagnation point: experiments and lattice Boltzmann simulations. *Phys. Rev. Lett.* 72: 3514–3518.

Taylor, G. I. (1953). Dispersion of a soluble matter in solvent flowing slowly through a tube. *Proc. R. Soc. Lond. A* 219: 186–203.

(Ref: Esch, 1960) Esch, H. (1960). Über die Körpertemperaturen und die Wärmehaushalt von *Apis mellifera. Zeitschrift vergleichende Physiologie* 43: 303–335.

(Ref: Lattice-Bolzmann Models)

Wolf-Gladrow, D. A. (2000). *Lattice-Gas Cellular Automata and Lattice Boltzmann Models: An Introduction* (*Lecture Notes in Mathematics 1725*). Springer, New York. (An introduction for graduate students. Chapter 5 discusses the properties of these models and a method for constructing them.)

LaBarbara, M. (1990). Principles of design of fluid transport systems in zoology, *Science* 249: 992–1000.

Palmer, B. J. (1998). Direct simulation of hydrodynamic relaxation in microchannels. *J. Chem. Phys.* 109: 196.

Chen, S. and G. D. Doolan (1998). Lattice Boltzmann method for fluid flows. *Annu. Rev. Fluid Mech.* 30: 329–364. (These authors give an overview of a parallel algorithm for simulating single and multiphased flows having complicated boundary conditions and multiphased interfaces, and they simulate reaction-diffusion systems.)

(Ref: Heating and Cooling)

Chappell, M. A. (1982). Temperature regulation of carpenter bee (*Xylocopa californica*) foraging in the Colorado Desert of Southern California. *Physiol. Zool.* 55: 267–280.

Cooper, P., W. M. Schaffer and S. L. Buchmann (1985). Temperature regulation of honey bees (*Apis mellifera*) foraging in the Sonoran Desert. *J. Exp. Biol.* 114: 115.

Heinrich, B. (1980). Mechanisms of body temperature regulation in honeybees, *Apis mellifera*. II. Regulation of thoracic temperature at high air temperatures. *J. Exp. Biol.* 85: 73–87.

An introductory book to these subjects clearly written for a general audience is Heinrich, B. (1996). *The Thermal Warriors: Strategies of Insect Survival.* Harvard University Press, Cambridge, Massachusetts.

(Ref: Walking Particles)

Slade, G. (1996). Random walks. *Am. Sci.* 84: 146–153.

Shlesinger, M. F. (1989). Levy flights: variations on a theme. *Physica D* 38: 304–309.

Lavenda, B. H. (1985). Brownian motion. *Sci. Am.* 252: 70–85.

Klafter, J., M. F. Shlesinger and G. Zumofen (1996). Beyond Brownian motion. *Phys. Today* 49: 33–39.

Einstein, A. (1926). In: Fürth, R. (ed.) *Investigations on the Theory of the Brownian Movement.* Methuen & Co. Ltd., London (English translation).

Von Smoluchowski, M. (1906). Zur kinetischen Theorie der Brownschen Molekularbewegung und der Suspensionen. *Ann. Physik* 21: 756.

(Ref: Levy Flight Dynamics)

McCrea, W. H. and F. J. W. Whipple (1940). Random paths in two and three dimensions. *Proc. R. Soc. Edinburgh* 60: 281–298.

Hayot, F. (1991). Lèvy walk in lattice-gas hydrodynamics. *Phys. Rev. A* 43: 806–810.

Shlesinger, M. F., B. J. West and J. Klafter (1987). Lévy dynamics of enhanced diffusion: application to turbulence. *Phys. Rev. Lett.* 58: 1100–1103. (These authors introduce a stochastic process called a Lévy walk that is a random walk having a non-local memory that is coupled in space and in time in a scaling fashion. Lévy walks enhance diffusion, i.e. diffusion that grows as t^α, $\alpha > 1$. Applying the idea of a Lévy walk to a particle diffusing in a fluid whose flow fluctuates, this model generalizes the idea of Taylor's correlated walk to give Richardson's t^3 law for the turbulent diffusion of a passive scalar in a Kolmogorov $-(5/3)$ homogeneous turbulent flow. The model also yields the deviations from the (5/3) exponent that Mandelbrot's intermittency gives. We may extend this model to describe chemical reactions occurring in turbulent flows.)

Viswanathan, G. M. *et al.* (1999). Optimizing the success of random searches. *Nature* 401: 911–914. For foraging when targets are sparse and can be visited any number of times, an inverse square power-law distribution of steps provides an optimal strategy.

(Ref: Jacobs and Hendrickson, 1997) Jacobs, D. and J. B. Hendrickson (1997). An algorithm for two-dimensional rigidity percolation: the pebble game. *J. Comput. Phys.* 137: 346–365. (Properties depend on the number of microscopic degrees of freedom. An algorithm is formulated as a pebble game and in the context of percolation through a rigid matrix, these authors calculate the degrees of freedom, the rigid clusters and locate the over-constrained regions.)

CHAPTER TEN: CONTROL

(Ref: Agarwal *et al.*, 2000) Agarwal, V., M. S. Hrishikesh, S. W. Keckler and D. Burger (2000). Clock rate versus IPC: the end of the road for conventional microarchitectures. In: *Proceedings of the 27th Annual International Symposium on Computer Architecture*.

(Ref: Chitin)

Cornell, J. C. and M. L. Pan (1983). The disappearance of molting fluid in the tobacco horn worm, *Manduca sexta*. *J. Exp. Biol.* 107: 501–504.

Caveny, S. (1991). Cell-to-cell signaling in the epidermis. In: Binningon, K. and A. Retnakaran (eds.) *Physiology of the Insect Epidermis.* CSIRO, Melbourne, pp. 25–35.

(Ref: Supercompilation)

Turchin, V. (1996). Supercompilation: techniques and results. In: *Perspectives of System Informatics*, *Lecture Notes in Computer Science*, Vol. 1181. Springer, Berlin.

(Ref: Network Flow Theory)

Ahuja, R. K., T. L. Magnanti and J. B. Orlin (1993). *Network Flows: Theory, Algorithms, and Applications*. Prentice Hall, New York. (A comprehensive introduction to shortest path, maximum flow, and minimum cost flow problems with descriptions of polynomial-time algorithms for these models as descriptions of important data structures, including d-heaps, Fibonacci heaps, and dynamic trees.)

Antoulas, A. C. (2005). *Approximation of Large-Scale Dynamical Systems.* SIAM Advances in Design and Control 6. (Book explains how to reduce complex mathematical models in view of limited computational accuracy and storage capabilities and the trade offs between accuracy and complexity.)

(Ref: Modeling, Rent's Rule, and Kolmogorov Complexity)

Rent's rule of a power-law scaling relation between the number of connections to a subsystem and its number of functional subunits within that subsystem applies both to engineered and biological systems having an exponent often between one half and one. Also physical constants are limiting (Gerschenfeld Ref., p. 162). Kolmogorov complexity put simply is that a good measure of the complexity of an object is the length of the shortest computer program that constructs the object.

Kolmogorov complexity is a versatile mathematical tool applicable for studying logical depth as well as the time and space complexity of computations. Kolmogorov complexity, also called algorithmic information or entropy, implies quantifying the amount of absolute information in an object as being point-wise or discrete rather than as the distributed classical average randomness produced by a random source. Traditionally we think that the better a model compresses the data from a phenomenon, the better we may learn, generalize and predict. To make these ideas rigorous means that we must find the length of the shortest effective description for the 'object' or its Kolmogorov complexity. Even though a best hypothesis does not necessarily permit a best prediction, compression appears to be one 'best strategy' for predictions. A comprehensive as well as introductory treatment at the graduate student level is the book by: Li, M. and P. M. B. Vitanyi (1997). *An Introduction to Kolmogorov Complexity and Its Applications.* Springer Verlag, New York. (For those with some background in computer science, the writing is very clear.)

(Ref: Chaos and Control)

Chen, G. and Y. Xinghuo (eds.) (2003). *Chaos Control: Theory and Applications* (*Lecture Notes in Control and Information Sciences*). Springer,

New York. (Chaos control theory for engineers containing information on liquid mixing, devices that include studies of the mammalian brain and heart.)

(Ref: Localization Within Hemocoels)

Kuhn, K. H., J. Uhlir and L. Grubhoffer (1996). Ultrastructural localization of a sialic acid-specific hemolymph lectin in the hemocytes and other tissues of the hard tick, *Ixodes ricinus* (Acari; Chelicerata) *Parisitol. Res.* 82: 215–221. (Lectins isolated from the hemolymph of most arthropods are molecules thought to function in invertebrate immunity. When labeled with antibodies for specific molecules, lectins injected into hemocoels revealed places in the hemocoel where the antibodies contacted receptors. In one case, a tick, the receptors were seen in granular hemocytes, immune cells of the hemolymph, as well as on the midgut and kidney.)

INDEX